Power Systems

Electrical power has been the technological foundation of industrial societies for many years. Although the systems designed to provide and apply electrical energy have reached a high degree of maturity, unforeseen problems are constantly encountered, necessitating the design of more efficient and reliable systems based on novel technologies. The book series Power Systems is aimed at providing detailed, accurate and sound technical information about these new developments in electrical power engineering. It includes topics on power generation, storage and transmission as well as electrical machines. The monographs and advanced textbooks in this series address researchers, lecturers, industrial engineers and senior students in electrical engineering.

More information about this series at http://www.springer.com/series/4622

Amal Souissi · Imen Abdennadher
Ahmed Masmoudi

Linear Synchronous Machines

Application to Sustainable Energy
and Mobility

Amal Souissi
Renewable Energies and Electric Vehicles
    Lab. (RELEV), Sfax Engineering
    National School (ENIS)
University of Sfax
Sfax
Tunisia

Ahmed Masmoudi
Renewable Energies and Electric Vehicles
    Lab. (RELEV), Sfax Engineering
    National School (ENIS)
University of Sfax
Sfax
Tunisia

Imen Abdennadher
Renewable Energies and Electric Vehicles
    Lab. (RELEV), Sfax Engineering
    National School (ENIS)
University of Sfax
Sfax
Tunisia

ISSN 1612-1287   ISSN 1860-4676   (electronic)
Power Systems
ISBN 978-981-13-4413-8   ISBN 978-981-13-0423-1   (eBook)
https://doi.org/10.1007/978-981-13-0423-1

Printed on acid-free paper

This Springer imprint is published by the registered company Springer Nature Singapore Pte Ltd.
part of Springer Nature
The registered company address is: 152 Beach Road, #21-01/04 Gateway East, Singapore 189721,
Singapore

Preface

Development that meets the needs of the present without compromising the ability of future generations to meet their own needs are the few words commonly expressed to define the so-called: sustainable development. It lies on the principle of continuously tracking the interference area shared by three leading requirements, that are

- increasing economical benefits,
- improving the quality of life,
- protecting the life cycle.

Within the fulfilment of the last requirement, much attention has been and continue to be given to the substitution of the polluting systems by eco-friendly ones. Such a substitution covers several vital fields. Of particular interest are energy and mobility.

With regard to energy, many research and development (R&D) projects were launched worldwide, with emphasis on the investigation of the power potential of classical and emergent earth's natural energy reserves. Much attention has been and continue to be paid to the wind and solar energies, which are presently considered as viable candidates to assist, in the short term, then substitute, in the long term, the carbon-based energy.

Beyond the conventional renewable energies, an increasing interest is presently given to the harvesting of emergent green sources exhibiting promising potentials, such as marine energies, covering tidal, marine current, temperature gradient, salinity and wave. This latter exhibits an energy density by far higher than all other forms of marine energy and even than the wind one, enabling dedicated conversion systems to harvest higher power. Among the most promising converters, those equipped with linear generators are particularly in harmony with the oscillating motion of the waves. They are considered as "direct drive" harvesters in so far as the required mechanisms that interface the wave oscillating motion to the machine shaft in the case of rotating generators turn to be useless.

Within the integration of linear generators in renewable and free energy harvesting, wave energy converters are by far the most significant power application, but it is not the sole. Indeed, considering smaller power range, linear generator could suitably equip several harvesters of free energies, such as

- human body motion energy,
- vibration-based energy.

Regarding mobility, until the 1960s, automotive manufacturers did not worry about the cost of fuel. They had never heard of air pollution and they never thought about life cycle. Ease of operation with reduced maintenance costs meant everything back then. In recent years, clean air policies are driving the market to embrace new propulsion systems in an attempt to substitute or to efficiently assist the internal combustion engine (ICE) by an electric drive unit, yielding the so-called electric or hybrid propulsion systems, respectively.

In the manner of road vehicles, railway ones are becoming more and more sustainable thanks to their improved energy efficiency. The most efficient railway transportation systems are without doubt MAGLEV trains. The high-energy efficiency is mainly gained thanks to the eradication of the friction within the contacts of train–rail during propulsion. Indeed, the MAGLEV technology enables the substitution of the mechanical contacts in conventional trains by magnetic interactions between the rail and cabins. This offers (i) the possibility to reach up to 600 km/h, (ii) the reduction of the maintenance frequency and cost thanks to the elimination of wheels and track wear, (iii) the minimization of noise and vibration which is in favour of the passengers comfort and (iv) the elimination of the risk of slipping that enhances the safety.

From a topological point of view, a MAGLEV train represents a long-stator linear motor where the rail behaves as a stator and the cabins play the role of the mover. Linear machines are suitably integrated in several other sustainable applications related to mobility, including but not limited to

- free piston engines,
- electromagnetic suspensions,
- Ropeless elevators.

Within this trendy topic, the book emphasizes the suitability of liner machines to equip sustainable applications related to energy harvesting and mobility. The manuscript is structured in four chapters:

- The first one deals with a state of the art related to the integration of linear machines in sustainable applications. Prior to doing so, the study is initiated by a review of the topological variety of linear machines with their classification according to the morphology and the AC type. Then, the selected sustainable applications equipped with linear machines are briefly described with emphasis on those applied to mobility and free and renewable energy harvesting.

- The second chapter is aimed at an analytical modelling of the electromagnetic forces exhibited by linear machines. The study is initiated by an overview of the electromagnetic phenomena. The electric and magnetic material properties and specifications are first recalled. A formulation of the magnetostatic and magnetodynamic models based on the main electromagnetic laws and the *Maxwell* equations is then carried out. The chapter is achieved by a prediction of the electromagnetic force, which is based on the relations between the energy and co-energy in the case of conservative electromagnetic systems.
- The third chapter treats one of the emergent sustainable applications, that is wave energy harvesting using appropriate converters. To start with, wave energy converters are classified according to the technology of their power take-off systems with emphasis on the topology of the integrated generators including rotating and linear topologies. In light of the previous survey, it clearly appears that linear PM synchronous machines are the most viable candidates, thanks to their high force density and energy efficiency at low speeds. Of particular interest is the inset PM (IPM) tubular topology which offers an increase of the energy efficiency and an intrinsic cancellation of the radial attractive forces. The modelling of the IPM tubular-linear synchronous machines (T-LSMs) is then treated considering the magnetic equivalent circuit (MEC). Two design approaches to minimize the end effect are proposed and their effectiveness is checked by finite element analysis. The chapter is achieved by an extension of the validity of the proposed model to the time-varying features by incorporating the mover position in the MEC.
- The fourth and last chapter is devoted to the investigation of the operation of magnetically levitated (MAGLEV) trains. To start with, a study statement is carried out with emphasis on an historical overview of the involved technology as well as on a multi-criteria classification of MAGLEV trains. Owing to the fact that the levitation is the key function of MAGLEV trains, a special attention is paid to their classification according to their suspension system. Within this statement, the electromagnetic suspension (EMS) and electrodynamic suspension (EDS) systems are deeply investigated. As regards the EMS technology, two electromagnet configurations are studied which are the flat track with U-shaped core electromagnet and the U-shaped track with U-shaped core electromagnet. Concerning the EDS technology, two systems are considered which are a moving SC coil over a conducting sheet and a moving SC coil facing a figure-eight null-flux coil.

Sfax, Tunisia

Dr. Amal Souissi
Assoc. Prof. Imen Abdennadher
Prof. Ahmed Masmoudi

Contents

About the Authors

Amal Souissi received the B.S. degree in electromechanical engineering in 2014, and the Ph.D. in electrical engineering in 2017, both from Sfax Engineering National School (SENS), University of Sfax, Sfax, Tunisia. She is a member of the Research Laboratory on Renewable Energies and Electric Vehicles (RELEV) of the University of Sfax. She published up to five journal papers, four among which in IEEE transactions and she presented up to six papers in international conferences. She has received the award of the best paper on renewable energy. She is a member of the publication committee of the International Conference on Ecological Vehicles and Renewable Energies, Monte-Carlo, Monaco. Her main interests include the design of new topologies of linear AC machines, applied to renewable energy harvesting as well as to MAGLEV transportation systems.

Imen Abdennadher received B.S. degree in electromechanical engineering, master's degree in electric machine analysis and control, Ph.D. and the research management ability degree in electrical engineering, all from Sfax Engineering National School (SENS), University of Sfax, Sfax, Tunisia, in 2005, 2006, 2012 and 2017, respectively. In 2009, she joined the Tunisian University, where she held different positions involved in both education and research activities. She is currently an Associate Professor of electric power engineering at SENS, a member of the scientific committee of the master on Sustainable Mobility Actuators: Research and Technology, and a member of the Research Laboratory on Renewable Energies and Electric Vehicles (RELEV), where she is the head of the machine design team. She published up to 15 journal papers, 7 among which in IEEE transactions. She presented up 22 papers in international conferences and 3 have been rewarded by the best presented paper prize. She is a member of the organizing committee of the International Conference on Ecological Vehicles and Renewable Energies, Monte-Carlo, Monaco. Her main interests include the design of new topologies of AC machines in motoring and generating modes, applied to renewable energy as well as to electrical automotive systems.

Ahmed Masmoudi received B.S. degree from Sfax Engineering National School (SENS), University of Sfax, Sfax, Tunisia, in 1984, Ph.D. from Pierre and Marie Curie University, Paris, France, in 1994, and the Research Management Ability degree from SENS, in 2001, all in electrical engineering. In August 1984, he joined Shlumberger as a Field Engineer. After this industrial experience, he joined the Tunisian University, where he held different positions involved in both education and research activities. He is currently a Professor of electric power engineering at SENS, the Head of the Research Laboratory on Renewable Energies and Electric Vehicles (RELEV) and the Coordinator of the master on Sustainable Mobility Actuators: Research and Technology. He published up to 85 journal papers, 19 among which in IEEE transactions. He presented up 367 papers in international conferences, 9 among which in plenary sessions, and 3 have been rewarded by the best presented paper prize. He is the co-inventor of an USA patent. He is the Chairman of the Program and Publication Committees of the International Conference on Ecological Vehicles and Renewable Energies (EVER), organized every year in Monte Carlo, Monaco, since 2006. He was also the Chairman of the Technical Program and Publication Committees of the first International Conference on Sustainable Mobility Applications, Renewables, and Technology (SMART) which has been held in Kuwait in November 2015. His involvement in the above conferences has been marked by an intensive guest-editorship activity with the publication of many special issues of several journals including the IEEE Transactions on Magnetics, COMPEL, ELECTROMOTION and ETEP. Professor Masmoudi is a senior member, IEEE. His main interests include the design of new topologies of AC machines allied to the implementation of advanced and efficient control strategies in drives and generators, applied to renewable energy as well as to electrical automotive systems.

List of Figures

List of Tables

Chapter 1
Linear Machines: State of the Art with Emphasis on Sustainable Applications

Abstract The chapter is aimed at a state of the art related to the integration of linear machines in sustainable applications. Prior to do so, the study is initiated by a review of the topological variety of linear machines with their classification according to the morphology and the AC-type. Then, selected sustainable applications equipped with linear machines are briefly described with emphasis on those applied to mobility and free and renewable energy harvesting. Regarding mobility, the selected applications are free piston engines, electromagnetic suspensions, MAGLEV trains, and ropeless elevators. Concerning energy harvesting applications, the selected applications are human body motion energy, vibration-based energy, and wave energy conversion.

Keywords Linear machines · Topological variety · Morphology/AC-type
Sustainable mobility · Free piston engines · MAGLEV trains · Ropeless elevators
Energy harvesting · Wave energy

1.1 Introduction

Linear machines have been and continue to be viable candidates in many industrial applications where they exhibit higher performance than their rotating counterparts. Of particular interest is the integration of linear machines in applications where the improvement of energy efficiency is a vital requirement. Indeed, they are systematically involved, as generators or motors, in applications incorporating linear motion. Thanks to the advances made in linear machine technology, mechanisms dedicated to the conversion of rotating motion into linear one and vice versa, such as gear boxes, turn to be useless, leading to the so-called "direct drive" linear electromechanical conversion.

Beyond the energy efficiency, linear machine concepts exhibit:

- high velocity which is no longer limited by the rotation-linear motion conversion in the case where rotating machines are considered,
- high acceleration obtained thanks to the direct mechanical coupling of the linear machine and its prime mover under generator operation and its load under motor

© Springer Nature Singapore Pte Ltd. 2019
A. Souissi et al., *Linear Synchronous Machines*, Power Systems,
https://doi.org/10.1007/978-981-13-0423-1_1

operation. While in the case of rotating machines, the rotation-linear motion conversion leads to an increase of the chain inertia which affects the electromechanical system dynamic,

- high accuracy of the position sensing. Whereas in the case of rotating machines, the position accuracy is affected by the backlash between the different driving element. Such a phenomenon is irregular and varies with respect to temperature and time,
- high lifetime with less maintenance. However when using rotating machines, the rotation-linear motion converters require a periodic maintenance and has a limited lifetime.

Despite the potentialities exhibited by linear machines, they present some limitations, among which one can distinguish the end effect phenomenon which is due to their open magnetic circuit. The end effect affects linear machine features. This drawback has been tackled in many works developed so far. Currently, there are several design approaches that enable a quasi-cancellation of the end effect.

Within this technological trend, the present chapter is aimed at a state of the art regarding the potential applications equipped with linear machines, with emphasis on the sustainable ones. Prior to do so, a topological variety of linear machines is established considering their classification according to the morphology and the AC-type.

1.2   Linear Machine Topological Variety

Linear machines have been and continue to be designed considering a large variety of topologies. They could be classified according to several criteria, especially the morphology and the AC-type. These are detailed here under.

1.2.1   Morphological Classification

Linear machines could be flat or tubular. They could have a long stator (the mover is shorter than the stator) or a short stator (the mover is longer than the stator). Moreover, the stator slots could be of single layer type or double layer one. A schematic diagram highlighting the previously-cited morphological classes of linear machines is provided in Fig. 1.1.

1.2.2   AC-Type Classification

Linear machines have been designed during many decades as squirrel cage induction actuators with specific designs of the mover "cage". Starting from the 1980s, the

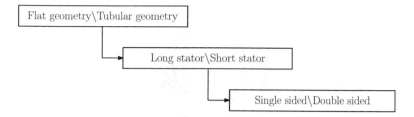

Fig. 1.1 Classification of linear machines according to their morphology

Fig. 1.2 Layout of a IPM
tubular-linear synchronous
machine

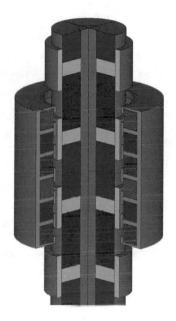

emergence of power electronic converters makes possible the development of linear
synchronous machines, as "converter-fed machines". These could be equipped with
a variable reluctance or a permanent magnet (PM) excitation in the mover.

PM-excited machines could be characterized by the circulation of the flux within
three dimensional (3D) paths (case of transverse flux machines) or two dimension
(2D) ones. Unlike rotating machines, there are limited PM arrangements in the mover.
The most considered ones are surface-mounted PMs (SPMs) and inset ones (IPMs).
Figure 1.2 shows the layout of a tubular-linear IPM synchronous machine.

Furthermore, a specific PM arrangement achieved according to the *Halbach* array
has been also considered. It exhibits attractive features, such as [1]:

1. a sinusoidal distribution of the air gap flux density which is desirable for low
   stator iron loss, sinusoidal back-EMF waveform (appropriate for brushless AC
   motors),

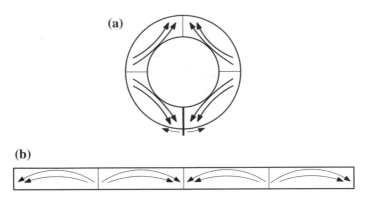

Fig. 1.3 *Halbach* array PM arrangement in the machine moving part (rotor/mover). **Legend:** **a** case of rotating machines, **b** case of linear machines

2. a self shielding magnetisation so that mover back-iron turns to be useless. The resulting mass reduction of the mover makes the machine interesting for high dynamic performance,
3. a low cogging force which makes the PMs skewing unnecessary to achieve a ripple-free force. This represents a crucial manufacturing cost benefit.

Figure 1.3 shows the principle of the PM magnetization within the moving part of *Halbach* array machines. It should be underlined that the lines separating the machines poles are fictitious.

Two distinguished topologies of *Halbach* concepts have been reported in the literature, which are: (i) the single ring *Halbach* array and (ii) the segmented *Halbach* array [2]. From a manufacturing point of view, the ring *Halbach* array requires a special magnetized device, while the segmented one could be simply achieved by installing magnetic blocks in a certain order that are magnetized in different directions. Further simplification of the manufacturing process, allied to an improvement of the cost-effectiveness, could be gained with the reduction of the number of the magnetic blocks [3, 4].

Of particular interest is the so-called "quasi-*Halbach*" concept where a pole is achieved by no more than three magnetic blocks: a radially-magnetized PM sandwiched in between two flux-concentrating arrangement ones, as illustrated in Fig. 1.4. The simplicity of such a *Halbach* concept makes it an attractive topology, fulfilling the tradeoff simplicity/performance.

Beyond the PM arrangements, linear synchronous machines could be classified according to the type of the armature winding, as follows:

• distributed winding with a slot per pole and per phase equal of higher than unity,
• concentrated winding with a slot per pole and per phase fractional lower than unity, yielding the so-called "fractional-slot PM synchronous machines" (FSPMSMs).

An increasing attention is presently given to FSPMSMs which is motivated by [5–9]:

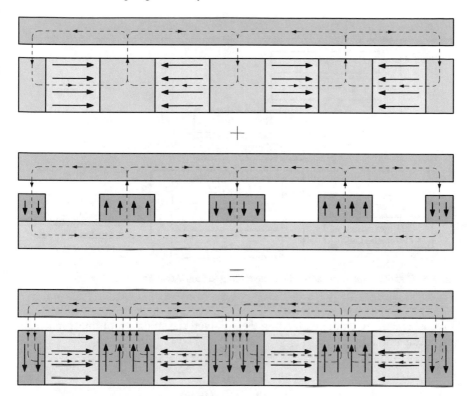

Fig. 1.4 Principle of achieving the quasi-*Halbach* PM arrangement based on the superposition of the surface mounted and the flux concentrating ones

- their low copper losses achieved thanks to their short end-windings,
- their reduced cogging torque,
- their high fault tolerance capability,
- their wide flux weakening range.

With this said, it has underlined that the above-listed performance are compromised by a dense harmonic content of the armature flux density spatial repartition, resulting in high eddy-current loss in the PMs. Such a drawback affects the machine energy efficiency and could lead to an irreversible demagnetization of the PMs. Several design approaches have been proposed to tackle the excessive eddy-current loss in the PMs. Referring to the literature, it has been reported that the most effective approach is the PM segmentation.

In [10], a 3D analytical model dedicated to the prediction of the PM eddy current loss of FSPMSMs, has been proposed and applied to the investigation of the effect of PM segmentation on the reduction of these loss, considering both radial and axial segmentations. The obtained results revealed the superiority of the radial segmentation.

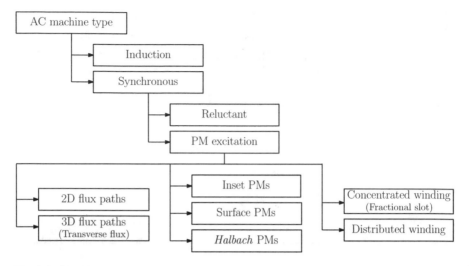

Fig. 1.5 Classification of linear machines according to their AC-type

A schematic diagram highlighting the AC-type classification of linear machines is illustrated in Fig. 1.5.

1.3 Integration in Sustainable Applications

1.3.1 Mobility Applications

1.3.1.1 Free Piston Engines

Basically, hybrid propulsion systems could be classified within two major families:

- the parallel hybrid propulsion systems: both internal combustion engine (ICE) and electric motor are involved in the propulsion of the vehicle. The denomination "parallel hybrid system" comes from the fact that the mechanical power driving the wheels flows within two parallel paths. The battery pack is charged by switching the electric machine operation from motor to generator. Although it has a simple structure, the parallel hybrid system is not able to drive the wheels by the electric motor while simultaneously charging the battery pack since the system is equipped with just one electric machine.
- the series hybrid propulsion systems: they are equipped with two electric machines, such that: (i) an electric motor used as a propeller, and (ii) a generator driven by the ICE. The generator is feeding the motor and charging the battery pack. The denomination "series hybrid system" comes from the fact that the power generated by the ICE flows in series through the generator and the propulsion motor to reach

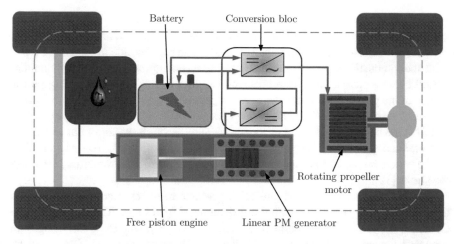

Fig. 1.6 Bloc diagram of a free piston engine-based series hybrid propulsion system equipped with a tubular-linear permanent magnet (PM) synchronous generator

the wheels. The ICE is operated at its rated speed in order to improve the energy efficiency and to reduce the emissions of green house gases.

Since the introduction of the hybrid propulsion systems in the beginning of the 1970s, several technological improvements have been brought to the field. For instance, some automotive manufacturers such as Toyota considered the series-parallel hybrid propulsion systems to equip their road vehicles. A series-parallel hybrid propulsion system consists in a combination of the series and parallel ones in order to maximize their benefits and consequently increase their energy efficiency.

Beyond the latter improvement, much interest is presently given to the power chain of series hybrid propulsion systems. Of particular interest is the ICE-generator topology which has been totaly rethought through the introduction of the so-called "free piston engine". Basically, it consists in a single or several combustion chambers/single piston to which is linked the mover of a linear generator. Thus, the conversion of the piston oscillating motion into rotation has been eliminated which represents crucial energy efficiency and compactness benefits.

A key component of the free piston engines is the linear generator that could be a tubular linear synchronous with inset PMs [11], or *halbach* array PMs [12, 13], or transverse flux [14], or tubular linear induction [15]. Figure 1.6 illustrates the block diagram of a free piston engine-based series hybrid propulsion system.

The concept of free piston engine has been introduced in 1925 by *Jordan* [16]. Starting from this earlier concept, several topologies have been successfully integrated in applications such as air compressors, gas generators and hydraulic pumps. In 1957, General Motors developed the first automotive free piston engine. It has been operated by diesel fuel using a pair of free piston gas generators engine.

Recently, Toyota Central R&D Labs., Inc. developed a free piston engine to be implemented in a hybrid electric vehicle manufactured by the same company [17].

It consists of a two-stroke combustion engine, a permanent magnet tubular-linear synchronous generator, and a gas spring chamber. The main feature of this design consists in a hollow circular step-shaped piston. The smaller side of the piston represents the combustion chamber, and the larger side constitutes the gas spring chamber. An oil cooling passage has been arranged in order to improve the cooling performance of the piston [18]. It has the merit to generate electricity with a high thermal efficiency reaching 42%; the ICE one does not exceed 30%. As far as the piston speed is null at the beginning and end of the stroke and reaches a maximum at its middle, it is preferable that the linear generator operation is restricted to the middle of the stroke which enables a free-movement of the piston during the remaining period of the stroke [19].

Several other free piston engine concepts have been reported in the literature, among which one can distinguish the project developed by the German Aerospace Center. It has led to the prototype shown in Fig. 1.7 [20]. Due to its flat design, the proposed concept could be integrated in the under-body of the car. Different variants including:

(a) single combustion chamber design,
(b) central gas spring design,
(c) central combustion design,
(d) central combustion design with integrated gas springs,
(e) central combustion design with branched linear generators,

have been proposed and compared.

Compared to series-hybrid propulsion systems equipped with a conventional ICE and a rotating generator, the free piston engine-based ones exhibit several advantages, such that:

- the powertrain from the piston to the generator is simplified because of the exclusion of the connecting rods, the crankshaft and the flywheel. The piston side loads are virtually removed which enables the elimination of the gearbox converting the linear motion of the piston into a rotating one. Thus, the lubrication system could be downsized or totally removed. All cited factors contribute to the engine efficiency by reducing its heat and frictional losses. Furthermore, the engine cost-effectiveness is improved,
- the piston motion can be adapted according to the fuel being used and is no longer constrained by the crankshaft to follow a fixed sinusoidal motion over time. This represents a crucial ecological benefit as far as biofuels could be used,
- the piston motion can be considerably extended allowing over-expansion of burned gases to extract the maximum possible work before these are released. Whereas in crank engines the piston's stroke is defined by the crankshaft. A further benefit, gained from such an over-expansion, results in the exhaust noise reduction.

In spite of the above-listed potentialities and up to date, free piston engines do not reach a sufficient technological maturity that enables their wide integration in automotive applications. Some limitations need to be investigated in depth, such as:

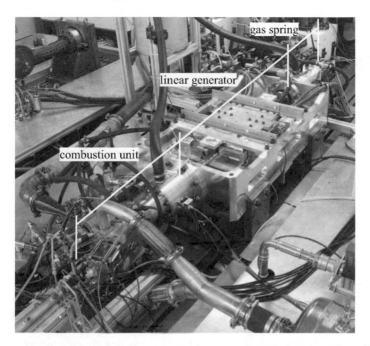

Fig. 1.7 Prototype of the free piston engine developed by the German Aerospace Center [20]

- in the absence of a crankshaft, multiple pistons must be accurately positioned and synchronised. If each piston's motion is not controlled adequately, the compression rate and ratio would vary leading to combustion variations and potential misfires. Both drawbacks could reduce the engine's efficiency and increase tailpipe emissions. If pistons are not synchronised, the engine would not be electrically and mechanically balanced, leading to harsh vibrations and power spikes,
- tailpipe emissions must be carefully controlled. Free piston engines typically operate a two stroke cycle. This latter can produce hydrocarbon emissions from unburned fuel and from inefficient scavenging processes. Whilst a free piston engine's high compression ratio delivers better efficiency with elevated temperatures. The pressures can lead to the formation of nitrous oxide: a harmful pollutant,
- some critical issues related to the tubular-linear permanent magnet (PM) synchronous machine design should be accounted for, such as:

  – the high temperature due to the combustion taking place in the engine chamber. A special attention must be paid to the cooling of the linear generator in order to prevent the damage of its components especially the irreversible demagnetization of the magnets,
  – the mechanical stresses due to the high oscillating frequency required to achieve the rated operating point of the ICE and therefore minimize the emissions of green house gazes,

– the high oscillating frequency could lead to significant iron loss. Hence, factional-slot concentrated windings in the armature should be avoided in so far as the PM eddy-current loss could be critical.

1.3.1.2   Electromagnetic Suspension Systems

Up to date, most road vehicles are equipped with passive hydraulic suspensions that consider hydraulic oil as a damper. Their role consists in the absorbtion of the excitation force caused by road irregularities. Nevertheless, these hydraulic systems present some drawbacks, such as:

• the environmental pollution due to hose leaks and ruptures of hydraulic fluids which are a source of toxicity,
• the lack of efficiency due to the required continuously-pressurized system.

Several attempts to tackle these limitations have led to the introduction of the electro-magnetic suspension systems which do not require an hydraulic fluid. Furthermore, they have a high force density that can efficiently-control the vehicle body vibration.

Despite the fact that the hydraulic and the electromagnetic suspension systems share a common role, they have several differences. The major one is related to the transformation of the absorbed energy. Indeed, the hydraulic suspensions transform the road excitation force into heat which is waisted. However, the electromagnetic suspension enables the harvesting of the absorbed energy and its conversion into electricity which is stored in the battery pack. The energy conversion is suitably-achieved thanks to a linear generator integrated in the electromagnetic suspension [21, 22].

Figure 1.8 illustrates the layouts of the conventional hydraulic and the electro-magnetic suspension systems. It clearly highlights their main difference, that is the co-axial tubular-linear PM synchronous generator equipping the electromagnetic suspension system.

The electromagnetic suspension systems could be classified as follows [22]:

• Passive electromagnetic suspension systems which consist in a mechanical spring and a linear generator. This latter acts as a damper by absorbing the kinetic energy that results from road excitation and converts it into electrical energy.
• Active electromagnetic suspension systems which require an extra system to con-trol the critical parameters such as thrust force, and damping force. Fully active electromagnetic suspension systems can completely eliminate the vibration affect-ing the vehicle body. However, they are high power consumers. It should be under-lined that the linear machines, integrated in the active electromagnetic suspension systems, have a combined motor-generator operation.

Considering the last type, and in order to minimize the energy consumption, space envelope, and costs, semi-active suspension configurations are currently investigated. These consist in a combination of an actuator with passive elements. Hence, they combine the advantages of passive and fully active suspension systems [23].

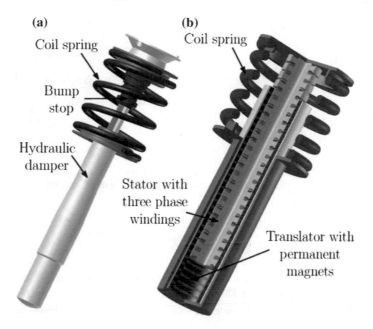

Fig. 1.8 Layouts of the two automotive suspension systems. **Legend: a** conventional hydraulic suspension system, **b** electromagnetic suspension system [21]

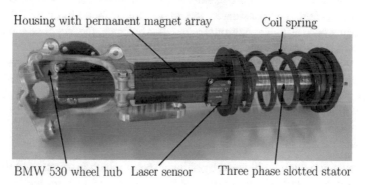

Fig. 1.9 Prototype of the electromagnetic active suspension system equipping the BMW 530i together with the wheel hub [25]

Electromagnetic suspension systems are equipped with linear machines having a tubular geometry to be in harmony with the application. These are commonly linear PM synchronous machines [21, 23–25], but could be also switched reluctance ones [26].

A prototype of the electromagnetic active suspension system equipping the BMW 530i together with the wheel hub are shown in Fig. 1.9 [25].

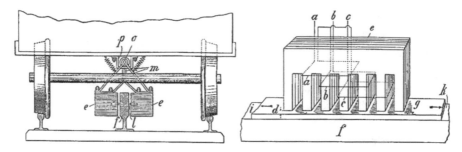

Fig. 1.10 First patented industrial transportation system built around a linear motor. **Legend**: (left) layout of the railway traction system, (right) layout of the induction motor used for the tain propulsion [4]

1.3.1.3   Transportation Applications

The first patent dealing with an industrial transportation application integrating a linear motor was published in 1902 in Germany by *Zehden* [27]. A linear induction motor has been selected to achieve the propulsion of a passenger train, as shown in Fig. 1.10.

Since then, many trains propelled by a linear motor have been developed. Linear motor-driven railway systems are usually adopted in mass transit systems for metro lines in city centers. The considered topologies are of induction type [28, 29], or flux switching one [30], or synchronous one. For instance, the Yokohama municipal subway train is equipped with a short stator linear induction motor.

The most interesting transportation application equipped with linear machines is without doubt the MAGLEV trains. They have been introduced in the 1970's and up to date they are considered as a state of the art topic. The most recent R&D project dealing with MAGLEV trains is the one developed in Japan [31].

The principle of operation of a MAGLEV train is based on three main functions which control the train movement on the guide-way, such that:

- the propulsion which is achieved by a linear motor which is usually fed through a stator winding installed in the guide-way. These motors could be induction [32–34] or synchronous [35–37] ones,
- the levitation which consists in keeping the train suspended without mechanical contact between the guide-way and the vehicle,
- the guidance which makes the vehicle following the guide-way.

The above-described functions are illustrated in Fig. 1.11.

Achieving the above-described MAGLEV functions simultaneously makes it possible the access to high degrees of performance, especially [38]:

(i) high speeds (up to 500 km/h) achieved by overcoming the main problem of high-speed railway trains, that is the mechanical contacts between the wheels and the rails,

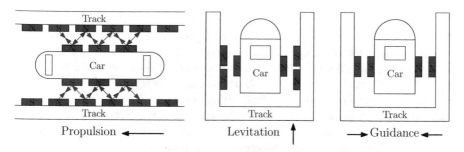

Fig. 1.11 Major functions on which lies the principle of operation of MAGLEV trains

(ii) low power consumption, less noise, more safety, and more comfort thanks to the absence of mechanical contacts,

(iii) high acceleration and deceleration capability.

Giving the significance of the MAGLEV technology which could be a serious competitor of the avionic one in the future, it will be treated in detail in Chap. 4.

1.3.1.4 Ropeless Elevators

The elevators equipped with steel rope and counter-weight, that we are familiar with today, were invented in 1854 opening the possibility to build tall skyscrapers and then save space for parks and open-space entertainment areas in highly populated cities. Considered as vertical mobility vehicles, they are also widely used in mines.

Since their introduction in the middle of the nineteenth century, elevators have been the subject of intensive technological advances. Following the hydraulically-propelled concepts, induction motors were used for single or double speed elevators where the car speed and passenger comfort were less of an issue. However, for higher performance and larger capacity elevators, the need for variable speed control turns to be necessary using DC motors powered by an AC/DC motor-generator. The latter topology enables a supply of the elevator controller separately from the rest of a building's electrical system. This offers the benefit of eliminating the transient power spikes in the building's electrical supply caused by the motors starting and stopping.

Starting from the 1980th, the widespread penetration of variable speed drives, in most if not all industrial fields, has enabled the integration of variable frequency static converter fed AC motors, especially of induction type, in the elevator traction chains. The steel ropes are attached to a hitch plate on top of the elevator car and then looped over the drive sheave to a counter-weight attached to the opposite end of the ropes. This mechanical concept is designed in such a way to reduce the power required for the car traction. The counter-weight is mounted in the hoistway and rides a separate railway system.

Recently, the need for elevators dedicated to very high skyscrapers has been emerged, in so far as they represents the predominant factor in the limitation of the buildings height. However, the technical requirements of such elevators could not be fulfilled by the conventional technology. Indeed, current elevators driven by rotating traction motors, would necessitate a 30% ratio of elevator hoistway space to the total floor space, which is obviously far from being economical. In addition and for the sake of safety consideration, the mass of the steel ropes would be very high leading to the increase of traction motor drive ratings which affects the elevator energy efficiency and cost-effectiveness. Moreover, the vertical vibration of very long ropes would be also a critical issue that would affect the control strategy performance. To these limitations are added the difficulties to drive more than two elevator cars per shaft. These can be only operated in the vertical direction.

These limitations could be totally overcome adopting the linear motor driven ropeless elevators. These represent viable candidates for very high skyscrapers.

From a topological point of view, the mover of the linear motor is mechanically-coupled to the elevator car through appropriate devices such as a suspension bar. The mover translates directly in front of the linear motor stator which is equipped with a three phase armature fed by a static converter under a closed loop control strategy. Ropeless elevators could be driven by flat geometry, PM or switched reluctance excitation, and single or double-sided (single of two air gaps) linear synchronous motors [39–42].

A basic ropeless elevator concept, equipped with a surface-mounted PM double sided linear synchronous motor, is illustrated in Fig. 1.12.

Referring to Fig. 1.12, one can notice the elimination of:

- the steel rope resulting in a theoretically-infinite propelling height. In all cases, thanks to the ropeless technology, the elevator is no longer limiting the hight of skyscrapers,
- the engine room located on the top of the elevator hoistway which is conventionally housing the rotating motor, the gearbox, the sheave, and the power electronic converter,
- the counter-weigh.

Beyond the basic concept, a multi-cars operation in one elevator hoistway can be conveniently-achieved by the ropeless technology, which greatly-reduce the elevator space occupancy. Furthermore, the ropeless technology makes it possible an horizontal transfer of the several cars of a given shaft to another one. This concept has been recently patent by *Otis Elevator Company* [43]. Indeed, an innovative exchange system allows the linear drive and guiding equipment to make 90°-turns of the car, by leveraging the linear motor technology developed for the MAGLEV trains. Hence, multiple cars travel safely up one shaft and down another in a single continuous loop, as shown in Fig. 1.13.

To sum up, it clearly appears that the ropeless technology has countered the limitations of the conventional one. However, a special attention has to be paid to the enhancement of the electromagnetic force. Indeed, with the elimination of the counter-weight, a high force density exhibited by the linear motor is of the most important feature for the ropeless elevator.

Fig. 1.12 Layout of a basic
ropeless elevator concept
equipped with a
surface-mounted PM double
sided linear synchronous
motor. **Legend**: (bottom)
face view, (top) top view

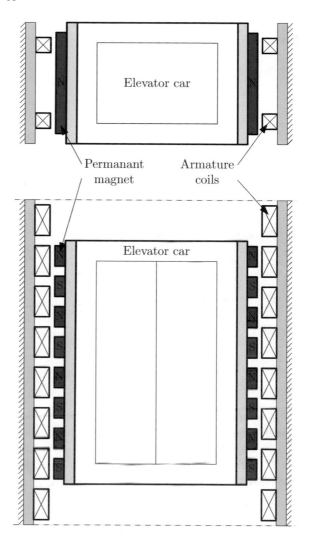

1.3.2 Energy Harvesting Applications

Following the alarming global warming phenomenon and the increasing power
demand, a renewed interest in green energies has emerged during the last two decades.
The penetration of these energies is conditioned by two major constraints:

- the policies and regulations regarding their interfacing to the traditional energy
  sources,
- the systems enabling their harvesting and distribution/storge systems.

Dealing with the energy harvesting systems, an increasing attention is currently
paid to take benefit of the potentialities of linear machines to convert free and

Fig. 1.13 Ropeless elevator
loop concept

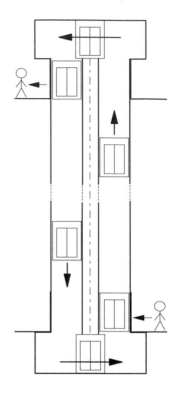

environment-friendly energies into electricity. Linear generator-based harvesting systems could be integrated in different applications, Covering a wide range of power, among which one can distinguish: (i) the human body motion energy, (ii) the vibration-based energy, and (ii) the wave energy. These are the subject of a brief description hereunder.

1.3.2.1 Human Body Motion Energy Harvesting

While moving, the human body is continuously producing a sustainable energy which could be harvested as:

- Backpack energy harvesting from human walking motion which takes advantage of the natural oscillation of the center of gravity. The displacement of the human body center varies from 3 to 7 cm at an average oscillation frequency of 1.95 Hz which changes according to the walking speeds. The harvesting system of this type of human body energy could be installed in a backpack. It could be equipped with flat linear generator with axially-magnetized PMs [44].
- Energy harvesting of the human motions of the different body parts, such as heel, ankle, knee, waist, elbow, wrist, and arm, as energy sources [45]. It has been shown that according to body parts, between 0.76 mW and 67 W could be harvested [46].

- Shoes energy harvesting system which consists in a small scale linear tubular PM generated fixed in the rear part of the shoe. During a walk, the linear generator mover has a variable velocity with respect to the stator [47]. This enables the generation of back-EMFs in the armature. These could be used, following an AC-DC conversion, to charge a small size embedded battery. Similar concept has been raised to power a flashlight, by shaking the linear generator with hand [48].

1.3.2.2 Vibration-Based Energy Harvesting

Vibration-based energy harvesting systems have been the focus of many recent investigations. This interest stems from the need to supply low-power consumption devices, such as micro electromechanical systems, health monitoring sensors, and wireless grid sensors. These harvesters can replace small batteries that can be used in aircraft systems, such as unmanned aircraft vehicles and micro air vehicles. These harvesters have also been proposed to power devices that rely on batteries, which have a finite-life span and are difficult or expensive to maintain. They can also be used to charge mobile phones. The transduction mechanisms used for transforming vibration to electric energy include: electromagnetic, electrostatic, and piezoelectric mechanisms.

Among the electromagnetic transducers, one can distinguish linear generators. These could be appropriately integrated in traffic energy harvesting systems which are installed on the road. This application has been considered in [49] where an optimization procedure has developed by means of hybrid evolutionary algorithms to reach the best overall system efficiency and the impact on the environment and transportation systems.

In [50], the modeling and optimization of a direct-drive contactless tubular linear generator have been developed in order to achieve a highly reliable device with long lifetime for vibration energy harvesting application. A speed bump in an urban environment has been treated as case study.

In [51], *Gros* et al. presented a comprehensive survey of energy harvesting concepts in the current social and technological context, with emphasis on linear generators as viable candidates in green energy conversion microsystems. The emerging technologies of energy harvesting are discussed, considering the environmental sustainable energy sources as well as the selection of the appropriate linear generators along with outlining aspects regarding their design and modeling. The survey also focused on the portability of the harvesting systems in an attempt to highlight the main advantages and disadvantages of these systems and to validate the challenges of using linear generators in this scope.

In [52], *Ye* et al. considered the power analysis of a single degree of freedom (DOF) vibration energy harvesting system built around a linear generator. The control strategy implemented in the linear generator has been designed in such a way to achieve a dual-role consisting in: (i) controlling the vibration and (ii) harvesting the energy. The power analysis of the studied DOF vibration energy harvesting system has highlighted that the generated electric power is greatly impacted by the

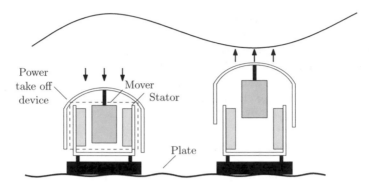

Fig. 1.14 *Archimedes* wave swing concept

dimensionless excitation frequency, the electric stiffness, and the electric damping coefficient/ratio of the linear generator.

1.3.2.3 Wave Energy Harvesting

Linear generator-based wave energy converters (WECs) are classified as direct drive systems. The wave oscillating movement is directly transmitted to the generator which represents a crucial energy efficiency benefit. The main linear generator-based WECs are described hereunder.

Archimedes **Wave Swing** (AWS) is a near-shore device which is the unique completely submerged WEC. This makes the system less vulnerable to storms. It consists in an air filled chamber which has the freedom to move in a vertical plane with respect to its base. When the wave passes over the AWS, a variation of the water depth causes a variation of the pressure surrounding the device which leads to its oscillation and consequently the one of the mover of the linear generator. AWS principle of operation is shown in Fig. 1.14. Referring to [53], it has been reported that the AWS has limited stroke, speed, and force.

Heaving Buoy (HB) consists in a floating body which could be cylindrical or spherical. The floating body follows the water surface in the vertical plane. It reacts against the seabed or submerged drag plate. The HB could be a near-shore or deep-water device. Its principe of operation is shown in Fig. 1.15.

   Several HB configurations have been reported in the literature. These are shown in Fig. 1.16. Much attention has been given to the configuration shown in Fig. 1.16a thanks to its simple structure.

   BH-based WECs have been the subject of many R&D projects among which one can distinguish:

- A 10 kW HB-based WEC was been developed in 2007 within the SeaBeavI project in USA. It is equipped with a tubular linear PM synchronous generator [54].

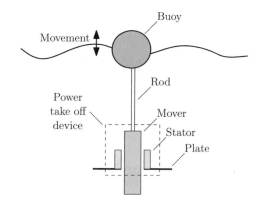

Fig. 1.15   Heaving buoy concept

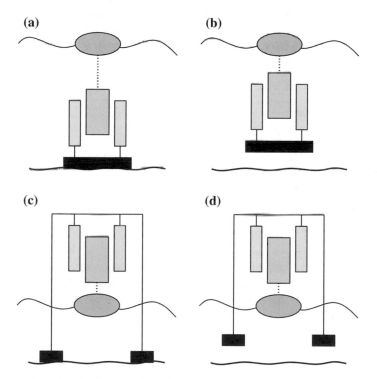

Fig. 1.16   Different heaving buoy configurations

- A 17 kW HB-based WEC developed within the project wave energy for a sustainable archipelago in 2012 in Åland (Finlande). The developed WEC is equipped with a tubular linear PM generator [55].
- A 10 kW HB-based WEC has been launched offshore outside the Swedish west coast in March 2006. It consists in a full-scale linear direct drive generator placed

on the seafloor. It is equipped with a four-sided linear PM flat generator which has a rated efficiency of 86% [56].

Accounting for the increasing attention given to WECs and their promising potentialities, they will be treated in depth in Chap. 3.

1.4 Conclusion

The first chapter dealt with a state of the art related to the applications integrating linear machines with emphasis on the sustainable ones. The study has been initiated by a review of the topological variety of linear machines, considering two classification criteria, such that: (i) the morphology, and (ii) the AC-type. Among the AC-types which are currently given an increasing attention, one can distinguish the linear PM synchronous machines which exhibit a wide variety according to the PM arrangement and the armature winding distribution. Then, a special attention has been paid to the integration of linear machines in selected sustainable applications which are:

- mobility applications, including:

  - free piston engines,
  - electromagnetic suspensions,
  - MAGLEV trains,
  - Repoless elevators.

- energy harvesting applications, including:

  - human body motion energy,
  - vibration-based energy,
  - wave energy.

Giving the increasing interest in wave energy harvesting and in MAGLEV trains, these will be treated in depth in Chaps. 3 and 4, respectively. Prior to do so, linear machine fundamentals aimed at their electromagnetic modeling and analysis are treated in the next chapter.

References

1. C. Xia, L. Guo, H. Wang, Modeling and analyzing of magnetic field of segmented halbach array permanent magnet machine considering gap between segments. IEEE Trans. Magn. **50**(12), 8106009 (2014)
2. K. Halbach, Design of permanent magnet multipole magnets with oriented rare earth cobalt material. J. Nucl. Instrum. Methods **169**(1), 1–10 (1980)
3. Y. Sui, P. Zheng, B. Yu, L. Cheng, J. Liu, Research on a tubular yokeless PM linear machine, in *Proceedings of the IEEE International Magnetics Conference*, Beijing, China, May 2015

4. M. Galea, L. Papini, H. Zhangand, C. Gerada, T. Hamiti, Demagnetization analysis for halbach array configurations in electrical machines. IEEE Trans. Magn. **51**(9), 810730 (2015)
5. A.M. EL-Refaie, T.M. Jahns, Optimal flux weakening in surface PM machines using fractional-slot concentrated windings. IEEE Trans. Ind. Appl. **41**(3), 790–800 (2005)
6. L. Alberti, M. Barcaro, N. Bianchi, Design of a low-torque-ripple fractional-slot interior permanent-magnet motor. IEEE Trans. Ind. Appl. **50**(3), 1801–1808 (2014)
7. I. Abdennadher, A. Masmoudi, Armature design of low-voltage FSPMSMs: an attempt to enhance the open-circuit fault tolerance capabilities. IEEE Trans. Ind. Appl. **51**(6), 4392–4403 (2015)
8. N. Bianchi, L. Alberti, M. Barcaro, Design and tests of a four-layer fractional-slot interior permanent-magnet motor. IEEE Trans. Ind. Appl. **52**(3), 2234–2240 (2016)
9. S.G. Min, G. Bramerdorfer, B. Sarlioglu, Analytical modeling and optimization for electromagnetic performances of fractional-slot PM brushless machines. IEEE Trans. Ind. Electron. **65**(5), 4017–4027 (2018)
10. A. Masmoudi, A. Masmoudi, 3-D analytical model with the end effect dedicated to the prediction of PM eddy-current loss in FSPMMs. IEEE Trans. Magn. **51**(4), 8103711 (2015)
11. P. Zheng, C. Tong, J. Bai, B. Yu, Y. Sui, W. Shi, Electromagnetic design and control strategy of an axially magnetized permanent-magnet linear alternator for free-piston stirling engines. IEEE Trans. Ind. Appl. **48**(6), 2230–2239 (2012)
12. J. Wang, M. West, D. Howe, H.Z.-D.L. Parra, W.M. Arshad, Design and experimental verification of a linear permanent magnet generator for a free-piston energy converter. IEEE Trans. Energy Convers. **22**(2), 299–306 (2007)
13. M.W. Zouaghi, I. Abdennadher, A. Masmoudi, No-load features of T-LSMs with quasi-Halbach magnets: application to free piston engines. IEEE Trans. Energy Convers. **31**(4), 1591–1600 (2016)
14. P. Zheng, C. Tong, G. Chen, R. Liu, Y. Sui, W. Shi, S. Cheng, Research on the magnetic characteristic of a novel transverse-flux PM linear machine used for freepiston energy converter. IEEE Trans. Magn. **47**(5), 1082–1085 (2011)
15. T.T. Dang, M. Ruellan, L. Prevond, H.B. Ahmed, B. Multon, Sizing optimization of tubular linear induction generator and its possible application in high acceleration free-piston stirling microcogeneration. IEEE Trans. Ind. Appl. **51**(5), 3716–3733 (2015)
16. E. Jordan, Generator of electric current, US Patent No. 1544010, June 1925
17. Toyota Central R&D Labs., Inc., Free piston engine linear generator "FPEG" (2014), www.tytlabs.com/tech/fpeg/index.html
18. H. Kosaka, T. Akita, K. Moriya, S. Goto, Y. Hotta, T. Umeno, K. Nakakita, Development of free piston engine linear generator system part 1-investigation of fundamental characteristics, SAE Technical Paper, No. 2014-01-1203 (2014)
19. S. Goto, K. Moriya, H. Kosaka, T. Akita, Y. Hotta, T. Umeno, K. Nakakita, Development of free piston engine linear generator system part 2-investigation of control system for generator, SAE Technical Paper, No. 2014-01-1193 (2014)
20. S. Schneider, F. Rinderknecht, H.E. Friedrich, Design of future concepts and variants of the free piston linear generator, in *Proceedings of the 2014 Ninth International Conference on Ecological Vehicles and Renewable Energies (EVER)*, Monte-carlo, Monaco, Mar 2014
21. B.L.J. Gysen, J.J.H. Paulides, J.L.G. Janssen, E.A. Lomonova, Active electromagnetic suspension system for improved vehicle dynamics. IEEE Trans. Veh. Technol. **59**(3), 1156–1163 (2010)
22. H.M. Isa, W.N.L. Mahadi, R. Ramli, M.A. Abidin, A review on electromagnetic suspension systems for passenger vehicle, in *Proceedings of the International Conference on Electrical, Control and Computer Engineering (INECCE)*, Kuantan, Malaysia, June 2011, pp. 399–403
23. J.J.H. Paulides, L. Encica, E.A. Lomonova, A.J.A. Vandenput, Design considerations for a semi-active electromagnetic suspension system. IEEE Trans. Magn. **42**(10), 3446–3448 (2006)
24. B.L.J. Gysen, J.L.G. Janssen, J.J.H. Paulides, E.A. Lomonova, Design aspects of an active electromagnetic suspension system for automotive applications. IEEE Trans. Ind. Appl. **45**(5), 1589–1597 (2009)

25. B.L.J. Gysen, T.P.J. van der Sande, J.J.H. Paulides, E.A. Lomonova, Efficiency of a regenerative direct-drive electromagnetic active suspension. IEEE Trans. Veh. Technol. **60**(4), 1384–1393 (2011)
26. J. Lin, K.W.E. Cheng, Z. Zhang, N.C. Cheung, X. Xue, Adaptive sliding mode technique-based electromagnetic suspension system with linear switched reluctance actuator. IET Electr. Power Appl. **9**(1), 50–59 (2015)
27. A. Zehden, Elektrische bef orderungsanlage unter benutzung eines wanderfeldmotors, German Patent no. 140958, June 1902
28. W. Xu, J.G. Zhu, Y. Zhang, Y. Li, Y. Wang, Y. Guo, An improved equivalent circuit model of a single-sided linear induction motor. IEEE Trans. Veh. Technol. **59**(5), 2277–2289 (2010)
29. J.-Q. Li, W.-L. Li, G.-Q. Deng, Z. Ming, Continuous-behavior and discretetime combined control for linear induction motor-based urban rail transit. IEEE Trans. Magn. **52**(7), 8500104 (2016)
30. R. Cao, M. Cheng, C. Mi, W. Hua, X. Wang, W. Zhao, Modeling of a complementary and modular linear flux-switching permanent magnet motor for urban rail transit applications. IEEE Trans. Energy Convers. **27**(2), 489–497 (2012)
31. H. Ohsaki, Superconducting Maglev—development and commercial service plan in Japan, in *Plenary session presented in the 2015 International Conference on Sustainable Mobility Applications, Renewables and Technology (SMART)*, Kuwait City, Kuwait, Nov 2015
32. Q. Lu, Y. Li, X. Shen, Y. Ye, Y. Fang, Y. He, Analysis of linear induction motor applied in low-speed maglev train, in *Proceedings of the International Conference on Electrical Machines and Systems (ICEMS)*, Sapporo, Japan, Oct 2012
33. Y. Guo, W. Xu, J. Zhu, H. Lu, Y. Wang, J. Jin, Design and analysis of a linear induction motor for a prototype HTS maglev transportation system, in *Proceedings of the International Conference on Applied Superconductivity and Electromagnetic Devices*, Chengdu, China, Sept 2009, pp. 81–84
34. Z. Deng, W. Zhang, J. Zheng, Y. Ren, D. Jiang, X. Zheng, J. Zhang, P. Gao, Q. Lin, B. Song, C. Deng, A high-temperature superconducting maglev ring test line developed in Chengdu, China. IEEE Trans. Appl. Supercond. **26**(6), 3602408(1–8) (2016)
35. H.-W. Cho, H.-K. Sung, S.-Y. Sung, D.-J. You, S.-M. Jang, Design and characteristic analysis on the short-stator linear synchronous motor for high-speed maglev propulsion. IEEE Trans. Magn. **44**(11), 4369–4372 (2008)
36. M.S. Hosseini, S. Vaez-Zadeh, Modeling and analysis of linear synchronous motors in high-speed maglev vehicles. IEEE Trans. Magn. **46**(7), 2656–2664 (2010)
37. J. Lee, J. Jo, Y. Han, C. Lee, Development of the linear synchronous motor propulsion testbed for super speed maglev, in *Proceedings of the International Conference on Electrical Machines and Systems (ICEMS)*, Busan, South Korea, Oct 2013, pp. 1936–1938
38. L. Yan, The linear motor powered transportation development and application in China. Proc. IEEE **97**(11), 1872–1880 (2009)
39. S. Masoudi, M.R. Feyzi, M.B.B. Sharifan, Force ripple and jerk minimisation in double sided linear switched reluctance motor used in elevator application. IET Electr. Power Appl. **10**(6), 508–516 (2016)
40. H.S. Lim, R. Krishnan, Ropeless elevator with linear switched reluctance motor drive actuation systems. IEEE Trans. Ind. Electr. **54**(4), 2209–2218 (2007)
41. S.-G. Lee, S.-A. Kim, S. Saha, Y.-W. Zhu, Y.-H. Cho, Optimal structure design for minimizing detent force of PMLSM for a ropeless elevator. IEEE Trans. Magn. **50**(1), 4001104 (2014)
42. X. Xu, X. Wang, S. Yuan, H. Feng, Optimization of vertical linear synchronous motor for rope-less elevator with INGA method, in *Proceedings of the International Conference on Electrical and Control Engineering (ICECE)*, Wuhan, China, June 2010, pp. 3965–3968
43. Z. Piech, T. Witczak, Ropeless elevator system, US Patent, Ref. US 2016/0297646 A1, Oct 2016
44. Z. Yang, A. Khaligh, A flat linear generator with axial magnetized permanent magnets with reduced accelerative force for backpack energy harvesting, in *Proceedings of the IEEE Twenty-Seventh Annual Applied Power Electronics Conference and Exposition (APEC)*, Florida, USA, Feb 2012, pp. 2534–2541

45. C. Ma, W. Zhao, L. Qu, Design optimization of a linear generator with dual halbach array for human motion energy harvesting, in *Proceedings of the IEEE International Electric Machines & Drives Conference (IEMDC)*, Idaho, USA, May 2015, pp. 703–708

46. P. Zeng, H. Chen, Z. Yang, A. Khaligh, Unconventional wearable energy harvesting from human horizontal foot motion, in *Proceedings of the 2011 Twenty-Sixth Annual IEEE Applied Power Electronics Conference and Exposition (APEC)*, Virginia, USA, Mar 2011, pp. 258–264

47. J.-X. Shen, C.-F. Wang, P.C.-K. Luk, D.-M. Miao, D. Shi, C. Xu, A shoe-equipped linear generator for energy harvesting. IEEE Trans. Ind. Appl. **49**(2), 990–996 (2013)

48. K. McCarthy, M. Bash, S. Pekarek, Design of an air-core linear generator drive for energy harvest applications, in *Proceedings of the 2008 Twenty-Third Annual IEEE Applied Power Electronics Conference and Exposition (APEC)*, Texas, USA, Feb 2008, pp. 1832–1838

49. A. Pirisi, M. Mussetta, F. Grimaccia, R.E. Zich, Novel speed-bump design and optimization for energy harvesting from traffic. IEEE Trans. Intell. Trans. Syst. **14**(4), 1983–1991 (2013)

50. L.A.J. Friedrich, J.J.H. Paulides, E.A. Lomonova, Modeling and optimization of a tubular generator for vibration energy harvesting application. IEEE Trans. Magn. **53**(11), 8209804(1–4) (2017)

51. I.-C. Gros, D.-C. Popa, P. Dorel Teodosescu, M. Radulescu, A survey on green energy harvesting applications using linear electric generators, in *Proceedings of the 2017 International Conference on Modern Power Systems (MPS)*, Cluj-Napoca, Romania, June 2017, pp. 1–5

52. J. Ye, Z. Lu, C. Chen, M. Wang, Power analysis of a single degree of freedom (DOF) vibration energy harvesting system considering controlled linear electric machines, in *Proceedings of the 2017 IEEE Transportation Electrification Conference and Expo (ITEC)*, Chicago, IL, USA, June 2017, pp. 158–163

53. H. Polinder, M.E.C. Damen, F. Gardner, Linear PM generator system for wave energy conversion in the AWS. IEEE Trans. Energy Convers. **19**(3), 583–589 (2004)

54. D. Elwood, S.C. Yim, J. Prudell, C. Stillinger, A. von Jouanee, T. Brekken, A. Brown, R. Paasch, Design, construction, and ocean testing of a taut-moored dual-body wave energy with a linear generator power take-off. Renew. Energy **35**(3), 348–354 (2010)

55. A. Savin, O. Svensson, M. Leijon, Research article study of the operation characteristics of a point absorbing direct driven permanent magnet linear generator deployed in the baltic sea. IET Renew. Power Gener. **10**(8), 1204–1210 (2016)

56. A. Savin, O. Svensson, M. Leijon, Estimation of stress in the inner framework structure of a single heaving buoy wave energy converter. IEEE J. Ocean. Eng. **37**(2), 309–317 (2012)

Chapter 2
Linear Machines: Electromagnetic Modelling and Analysis

Abstract The second chapter is devoted to the analytical modelling of the electromagnetic phenomena exhibited in linear machines. The study is initiated by an overview of electromagnetic basis. To do so, electric and magnetic material properties and specifications are firstly recalled. A formulation of magnetostatic and magnetodynamic models based on the main electromagnetic laws and the Maxwell equations is then carried out. Considering the case of conservative electromagnetic systems, a survey of the energy and the co-energy relations is provided. The prediction of the electromagnetic forces, related to the energy relations and based on the established models, is finally treated.

Keywords Linear machines · Analytical modelling · *Maxwell* equations
Magnetic vector potential · Energy and co-energy · Electromagnetic forces

2.1 Introduction

Thanks to the advances made in linear machine technology, mechanisms dedicated to the conversion of rotating motion into linear one and vice versa, such as gear boxes, turn to be useless, leading to the so-called "direct drive" linear electromechanical conversion.

Linear actuators could be classified considering the nature of the produced force, such as electromagnetic, electrostatic, piezoelectric, and magnetostriction forces. These devices are capable of producing direct drive motion. Such a motion could be a progressive (always same direction) or oscillatory. The most attention in the present book is paid to electromagnetic devices as they provide higher force density (N/kg) and higher efficiency (lower loss per force ratio (W/N)).

The design, sizing and optimization of electromechanical actuators require dedicated tools enabling the prediction of their magnetic and electric features. Basically, these tools consist in a set of equations, yielding the so-called model, that enable the emulation, as accurately as possible, of the behavior of the electromechanical actuators under different operating modes.

© Springer Nature Singapore Pte Ltd. 2019
A. Souissi et al., *Linear Synchronous Machines*, Power Systems,
https://doi.org/10.1007/978-981-13-0423-1_2

The absence of mechanical transmission in direct drive actuators opens up the possibility of providing a set of new topologies depending on the targeted application. To do so, solid fundamental knowledge on fields, forces, materials, and methods of the adopted approaches in the modeling and designing process should be focused in an earlier step.

Dealing with the modeling of the electromechanical actuators, there are different modeling approaches, among which one can distinguish:

- the *Maxwell* equations which are systematically solved using the finite element method in order to achieve high accuracies,
- the magnetic equivalent circuit, also named lumped circuit, that leads to lower accuracies than the previous model. However, it requires lower computation time which represents a crucial computer aided-design benefit,
- the analytical approaches based on the formulation of the flux density spatial repartition. These require lower CPU-time than the second model with more or less accurate results depending on the considered assumptions.

Considering the two last approaches, the present chapter deals with the analytical modeling of electromagnetic force exhibited by a linear actuators based either on the magnetic vector potential formulation in the air gap region, deduced from *Maxwell* equations or on the formulation of the magnetic energy and co-energy.

2.2 Electromagnetic Basis

The electromagnetic induction is the scientific principle that underlines several modern technologies, from the generation of electricity to communications and data storage.

More fundamentally, electromagnetic induction establishes an important link between electricity and magnetism, a link with important implications for understanding light as an electromagnetic wave.

The investigation of such electromagnetic phenomena goes through the identification of the involved electric and magnetic fields.

2.2.1 Electric Field

Let us define the electric field by considering the action of a given charge Q_1 placed at a given distance of a second charge Q_2. One can affirm that Q_1 creates an electric field $\vec{E}$ which exerts on Q_2 a force expressed as [1]:

$$\vec{F_E} = q\,\vec{E} \qquad\qquad (2.1)$$

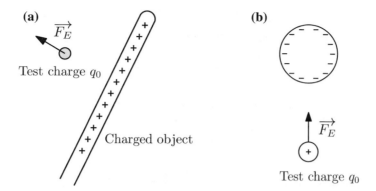

Fig. 2.1 Electric force applied to a test charge q_0. **Legend: a** existence of the electric force, and **b** electric force orientation

where $\overrightarrow{F_E}$ is the electric force and q is the electric charge of Q_2. Therefore the field $\overrightarrow{E}$ is defined as the force that would be felt by a unit positive test charge (q_0), as illustrated in Fig. 2.1a.

Note: The SI units for the electric field: newtons per coulomb (N/C).

The electric field lines could be visualized by drawing field lines which are defined by three properties, such as (Fig. 2.1b):

- Lines point in the same direction as the field,
- Density of lines gives the magnitude of the field,
- Lines begin on + charges and end on − charges.

2.2.2 Magnetic Field

Let us consider an uniform magnetic field passing through a surface S, as shown in (Fig. 2.2):

Fig. 2.2 Magnetic field passing through a surface S

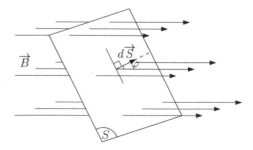

Let us define the area vector $\vec{S} = S d\vec{S}$, where S is the area of the surface and $d\vec{S}$ its unit normal. The magnetic flux through the surface is given by:

$$\Phi = \vec{B}\,\vec{S} = BS\cos\varphi \qquad (2.2)$$

When considering a non uniform distribution of the field density $\vec{B}$, Eq. (2.2) turns to be:

$$\Phi = \iint_S \vec{B}\,d\vec{S} \qquad (2.3)$$

Note: The SI units for magnetic flux is the weber (Wb): $1\,Wb = 1\,T\cdot m^2$

Based on the output force that they produce, the magnetic field can be defined with respect to the torque it produces on a magnetic dipole. Indeed, considering two separated magnets, one fixed on the ground and the second having the possibility of rotation around its center, if the two poles of two different magnets of similar polarity are placed near each other, the one having the possibility of rotation will turn in order to align itself with the first magnet. This magnetic torque tends to align a magnet's poles with the magnetic field lines.

Attractive force is generated between two magnetized objects of different polarity, while repulsive ones are caused by the field lines of two poles of opposite polarity facing each other.

The fundamental source of a magnetic field is a moving electric charge. Thus, a charge q, moving at a velocity of $\vec{u}$, produces a magnetic field around it. Based on the *Biot-Savart* law, the magnetic field $\vec{B}$ is defined as [1]:

$$\vec{B} = \frac{\mu}{4\pi}\frac{q\,\vec{u}\times(\vec{r}/r)}{r^2} \qquad (2.4)$$

where $\vec{r}$ is a vector between the charge and the point where the magnetic field is calculated oriented from the charge to the magnetic field.

The considered charge is then subject to a magnetic force $\vec{F}_B$ expressed as:

$$\vec{F}_B = q.(\vec{u}\times\vec{B}) \qquad (2.5)$$

Basically, a current I is the rate of charge-flow in a conductor moving along an elementary distance dl during dt as $I = q/dt$. Doing so, a magnetic field $\vec{B}$ could be produced by current carrying conductor as:

$$\vec{B} = \frac{\mu}{4\pi}\int\frac{I\,\vec{dl}\times(\vec{r}/r)}{r^2} \qquad (2.6)$$

Basically, when a conductor carrying an electric current I is placed in a magnetic field $\vec{B}$, each of the moving charges representing the current, experiences the *Lorentz* force. Considering the totality of the moving charges, one can evaluate a macro-force on the wire (*Laplace* force) which is expressed as [1]:

$$\overrightarrow{F}_B = \int I \overrightarrow{\Delta l} \times \overrightarrow{B} \tag{2.7}$$

Equation (2.7) could be used only to determine the force value. However, the force direction exerted on a conductor carrying a current $\overrightarrow{I}$ placed in a magnetic field $\overrightarrow{B}$ can be deduced using the right hand rule.

2.2.3 Magnetic and Electric Properties

The magnetic and electric properties of the considered materials are presented in what follows. Indeed, electric phenomena are resulting in the existence and the circulation of electric charges. The *Ohm*'s law, for electric conductors of electrical conductivity σ (S/m), is expressed as [2]:

$$\overrightarrow{J} = \sigma \overrightarrow{E} \tag{2.8}$$

In fact, in electrostatic problem the considered charge is supposed immobile. For the general case one can suppose that the charged wire is moving with a speed $\overrightarrow{u}$ with respect to the stationary reference system, in the presence of the magnetic field $\overrightarrow{B}$.

Under such hypothesis, the *Ohm*'s law (Eq. 2.8) could be rewritten as:

$$\begin{aligned} \overrightarrow{J} &= \sigma(\overrightarrow{E} + \overrightarrow{u} \times \overrightarrow{B}) \\ \overrightarrow{J} &= \sigma \overrightarrow{E'} \end{aligned} \tag{2.9}$$

with $\overrightarrow{E'} = \overrightarrow{E} + \overrightarrow{u} \times \overrightarrow{B}$ is the electric field with respect to the moving coordinate system.

In an other hand, magnetic phenomena are affected by the considered material. Depending on the material permeability μ, one can express the magnetic field density $\overrightarrow{B}$ as a function of the magnetic fields intensity $\overrightarrow{H}$ (A/m), so as [2]:

$$\overrightarrow{B} = \mu \overrightarrow{H} \tag{2.10}$$

where μ is the permeability and it is expressed in terms of the relative permeability μ_r of the considered materiel as:

$$\mu = \mu_r \mu_0 \tag{2.11}$$

where μ_0 is the absolute permeability equal to $4\pi 10^{-7}$.

Actually, magnetic materials are subject to two different types of non linearity leading to a limitation of the use of such material and enhance the modeling

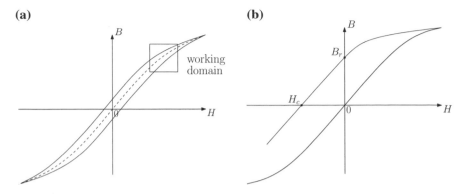

Fig. 2.3 (B-H) characteristics. **Legend: a** ferromagnetic material, and **b** permanent magnet material

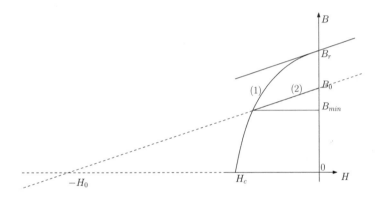

Fig. 2.4 Approximation of the demagnetizing characteristic of a permanent magnet material

difficulties. The non linearities are mainly the saturation and the hysteresis effects. Figure 2.3a shows the working domain of such non-linear phenomena [3].

Dealing with permanent magnet material, it is to be noted that a PM is mainly characterized by its magnetic curve and its geometry. Figure 2.3b shows the magnetic characteristic of the PM material within the *B-H* curve.

Moreover, the magnetic characteristic is delimited by two hot points, such as the remanent flux density B_r ($H = 0$) and the coercive magnetic field H_c (negative value of H_c for $B = 0$) (Fig. 2.4).

Starting from a point of the main (B-H) characteristic, one can figure out that the working point leaves this characteristic as soon as the induction increases. The corresponding curve can be approximated sufficiently by a straight line (2) which is defined between the demagnetizing characteristic (1) and the vertical axis, as illustrated in Fig. 2.4 [3]. This line is approximately parallel to the tangent of the demagnetizing characteristic at point M(0, B_r).

This line represents a linearized domain of the working point locus corresponding to a decrease of the flux density, until the operating point rejoins the main charac-

Fig. 2.5 Orientation of charged particles in a dielectric material. **Legend:** **a** non polarized condition, **b** polarized condition

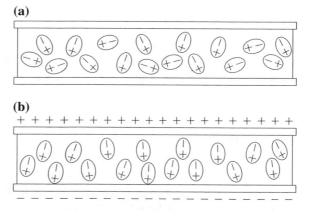

teristic curve. The limit is defined by the lowest point reached on the demagnetizing characteristic (B_{min}).

Under the operating conditions defined above, the linearized domain is represented by straight line which is characterized by a x-axis component $-H_0$ at the vertical axis origin and a y-axis component B_0 at the horizontal axis origin. The curve slope μ_d is defined as a permeability of the equivalent linear model, leading to the following working point induction B expression with respect to the magnetic field H:

$$B = B_0 + \mu_d H \qquad (\mu_d = \frac{B_0}{H_0}) \tag{2.12}$$

In electromagnetism, the electric displacement field $\vec{D}$ represents how an electric field $\vec{E}$ influences the organization of electric charges in a given medium domain, including charge migration and electric dipole reorientation. Its relation to permittivity In the very simple case of linear, homogeneous, isotropic materials with "instantaneous" response to changes in electric field is [2]:

$$\vec{D} = \varepsilon \vec{E} \tag{2.13}$$

where ε is the material permittivity.

In general, permittivity is not a constant, as it can vary with the position in the medium, the frequency of the field applied, humidity, temperature, and other parameters. In a nonlinear medium material sandwiched between the two polarized plates, the permittivity can depend on the strength of the electric field.

Figure 2.5 illustrates a dielectric material, under polarized(a) and non polarized(b) conditions, showing the orientation of charged particles. Such a material could be characterized by a lower ratio of electric flux to charge (more permittivity) than empty space.

Note: Electric conductivity σ, permittivity ε, permeability μ are scalars in isotropic materials, but they are tensor in anisotropic ones.

2.3 Electromagnetic Laws

Electromagnetic formulation is inevitably related to electric and magnetic phenomenon. To understand the fundamentals basis of such problems it is necessary to present a survey on the principal involved laws.

2.3.1 Gauss' Law

Gauss's Law is used to define the relation between the resultant electric field $\vec{E}$ strength and the total charge Q [4]. Actually, the net electric flux through any closed surface is proportional to the enclosed electric charge:

$$Q = \oint \varepsilon \vec{E}\, \vec{dS} = \oint \vec{D}\, \vec{dS} \tag{2.14}$$

where Q is the enclosed charge in the volume delimited by the considered surface, with:

$$Q = \int_v \rho dv \tag{2.15}$$

where ρ is the volemic conductivity in the considered material.

The electric flux density $\vec{D}$ is defined at area element $d\,\vec{S}$ of closed surface S. $d\,\vec{S}$ is a normal vector to plane of an elementary portion of the closed surface S (Fig. 2.6).

Based on the *Divergence* theorem, Eq. (2.14) leads to:

$$Q = \int_v \nabla.\vec{D}\, dv \tag{2.16}$$

which yields:

$$\nabla.\vec{D} = \rho \tag{2.17}$$

Fig. 2.6 Electric flux density over a Gaussian surface

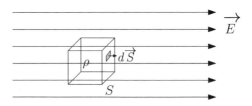

Fig. 2.7 Electric potential $\mathcal{V}(M)$ of the per unit charge q placed at the point M

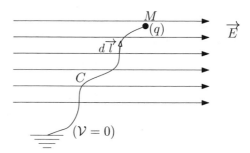

Fig. 2.8 Difference in electric potential between two points A and B

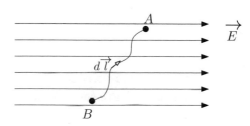

2.3.2 Electric Potential

The electric potential energy per unit charge depends on the position of the point M in which the charge is placed. Indeed, the electric potential of the per unit charge q, placed at the point M, corresponds to the amount of work that has to be done to move a point charge from a place of zero potential to M (Fig. 2.7).

The electric potential of the point M is given by [4]:

$$\mathcal{V}(M) = -\int_C \vec{E} \, \vec{dl} \tag{2.18}$$

where C is an arbitrary curve which connects a point of zero potential to the point M.

The difference in electric potential between two points A and B is the most used form of the electric potential energy, and it corresponds to the voltage drop between A and B (Fig. 2.8).

$$\mathcal{V}(A) - \mathcal{V}(B) = -\int_B^A \vec{E} \, \vec{dl} \tag{2.19}$$

2.3.3 Faraday's Law

The integral formulation of Eq. (2.19) can be converted into a differential form, which is applied under slightly different conditions. Doing so and considering an

Fig. 2.9 Induced magnetic field density under an electric field circulation

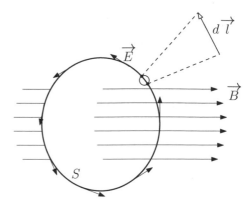

electric field flowing in a closed contour C, as shown in Fig. 2.9, the electric potential difference dV between two points separated by the distance $\vec{dl}$ is [5]:

$$d\,V = -\vec{E}\,\vec{dl} \tag{2.20}$$

with the integral form, one can write:

$$\mathcal{E} = V = \oint_C \vec{E}\,\vec{dl} \tag{2.21}$$

Considering the resultant magnetic field density $\vec{B}$ passing through a surface S, the total magnetic flux ϕ through S is defined as [5]:

$$\phi = \iint_S \vec{B}\,\vec{dS} \tag{2.22}$$

The obtained magnetic flux ϕ is time dependent and based on *Faraday*'s law one can write:

$$\mathcal{E} = -n\frac{d\varphi}{dt} = -\frac{d\phi}{dt} \tag{2.23}$$

where $\mathcal{E}$ is the electromotive force (EMF), which corresponds to the voltage generated around a closed loop, and φ is the magnetic flux per turns, n is the number of turn, and ϕ is the total magnetic flux. This latter could be expressed as the product of the area times the magnetic field normal to that area. Such a definition of magnetic flux makes the magnetic field density $\vec{B}$ often referred as magnetic flux density.

The negative sign in Eq. (2.23) is related to the *Lenz's law*. Indeed, the induced current in the wire resulting from the changing magnetic field, will create an opposite magnetic field. This latter is generated in an attempt to reduce or eradicate the cause that makes it existent.

Referring to Eqs. (2.21) and (2.23), one can write:

$$\oint \vec{E} \ \vec{dl} = -\frac{d}{dt} \iint_S \vec{B} \ \vec{dS} \tag{2.24}$$

Using *Stokes* theorem for vector fields, one can write:

$$\oint_\Gamma \vec{E} \ \vec{dl} = \iint_S (\nabla \times \vec{E}) \ \vec{dS} \tag{2.25}$$

The *Maxwell-Faraday* equation is a generalisation of *Faraday*'s law that states that a time-varying magnetic field is always accompanied by a spatially-varying, non-conservative electric field, and vice versa. The *Maxwell Faraday* equation is:

$$\nabla \times \vec{E} = -\frac{\partial \vec{B}}{\partial t} \tag{2.26}$$

Vector algebra shows that the divergence of a rotor field is zero:

$$\nabla.(\nabla \times \vec{E}) = 0 \tag{2.27}$$

Considering Eqs. (2.25) and (2.26), one can conclude:

$$\nabla.\vec{B} = 0 \tag{2.28}$$

Dealing with linear machines, the magnetic flux density vector $\vec{B}$ is always expressed in the general cartesian frame, such as:

$$\vec{B} = B_x \vec{x} + B_y \vec{y} + B_z \vec{z} \tag{2.29}$$

With this said, Eq. (2.28) turns to be:

$$\nabla.\vec{B} = \frac{\partial B_x}{\partial x} + \frac{\partial B_y}{\partial y} + \frac{\partial B_z}{\partial z} \tag{2.30}$$

Based on the Eq. (2.9) and taking into consideration a relative motion between the wire and the magnetic field density, *Maxwell Faraday* equation (2.26) should be updated such as:

$$\nabla \times \vec{E} = -\frac{\partial \vec{B}}{\partial t} + \nabla \times (\vec{u} \times \vec{B}) \tag{2.31}$$

Fig. 2.10 Induced magnetic
excitation field under an
electric current circulation

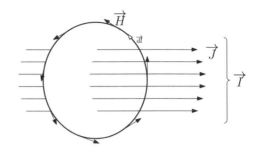

Fig. 2.11 Magnetic flux
tube

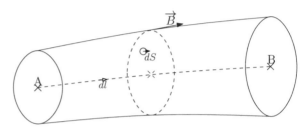

2.3.4 Ampere' Law

The *Faraday*'s law considers the relationship between the electric field $\overrightarrow{E}$ and the
resulting magnetic field density $\overrightarrow{B}$. While the *Ampere*'s law considers the relationship between existing (given) electric currents and the resulting magnetic fields $\overrightarrow{H}$.
Indeed, the *Ampere*'s law could be expressed in terms of $\overrightarrow{B}$ as [5]:

$$\oint \overrightarrow{H}\,\overrightarrow{dl} = \overrightarrow{I} \quad \Leftrightarrow \quad \oint \overrightarrow{B}\,\overrightarrow{dl} = \frac{1}{\mu}\overrightarrow{I} \tag{2.32}$$

where μ is the material permeability μ (H/m), and $\overrightarrow{I}$ is the electric current enclosed
by the closed path of integration as can be described in the Fig. 2.10.

The *Ampere*'s law could be rewritten as:

$$\oint \overrightarrow{H}\,\overrightarrow{dl} = \iint_{S} \overrightarrow{J}\,\overrightarrow{dS} = \overrightarrow{I} \tag{2.33}$$

where $\overrightarrow{J}$ is the surface current density (A/m^2) through the surface $\overrightarrow{S}$.

Again, *Stokes*' theorem Eq. (2.33) leads to:

$$\nabla \times \overrightarrow{H} = \overrightarrow{J} \tag{2.34}$$

In the other hand, the magnetic potential ξ_A^B along the flux line between the two
positions A and B, as illustrated in Fig. 2.11, is expressed based on the circulation

of the magnetic field along a closed path (Eq. 2.33) [3]:

$$\xi_A^B = \int_A^B \overrightarrow{H} \, \overrightarrow{dl} \tag{2.35}$$

Considering the relation between the magnetic field H and the flux density B and taking into account the collinearity between phasors $\overrightarrow{H}$ and $\overrightarrow{dl}$, Eq. (2.35) can be rewritten as follows:

$$\xi_A^B = \int_A^B \frac{B dl}{\mu} = \int_A^B B S_A^B \frac{dl}{\mu S_A^B} = \int_A^B \phi \frac{dl}{\mu S_A^B} \tag{2.36}$$

where S_A^B is the flux tube equivalent surface.

Taking into account the flux conservation, Eq. (2.36) turns to be as follows:

$$\xi_A^B = \phi \int_A^B \frac{dl}{\mu S_A^B} \tag{2.37}$$

Equation (2.37) yields the magnetic *Ohm* law which enables the definition of the reluctance $\mathcal{R}_A^B$ as the ratio of the magnetic potential $\mathcal{F}_A^B$ over the flux ϕ, as follows [3]:

$$\mathcal{R}_A^B = \frac{\xi_A^B}{\phi} = \int_A^B \frac{dl}{\mu S_A^B} = \frac{L_B^A}{\mu S_A^B} \tag{2.38}$$

where L_B^A is the flux tube mean length.

The validity of the above-developed formulation is restricted to the case of a linear behavior of the magnetic circuit as well as to the case of a relative magnetic permeability nearly or equal to unity.

Under saturated magnetic circuit, the reluctances corresponding to the iron parts of the magnetic circuit turn to be dependent on the corresponding fluxes. Their derivation is based on the *Hopkinson* law.

For instance, let us assume that the flux tube shown in Fig. 2.11 is located in saturated iron-made magnetic circuit. Then, the reluctance $\mathcal{R}_A^B$ turns to be dependant of the flux ϕ as follows:

$$\mathcal{R}_A^B(\phi) = \frac{L_A^B}{\phi} H \left(\frac{\phi}{S_A^B} \right) \tag{2.39}$$

where $H \left(\frac{\phi}{S_\phi} \right)$ is the magnetic field corresponding to the MEC-predicted flux, using the $B(H)$ characteristic of the involved iron material.

Moreover, the surface current density $\overrightarrow{J}$ is related to the volume charge density ρ by the continuity equation:

$$\nabla \cdot \overrightarrow{J} = -\frac{\partial \rho}{\partial t} \tag{2.40}$$

It should be underlined that there is an inconsistency between Eqs. (2.34) and (2.40), as follows:

$$\begin{aligned}
\nabla \cdot (\vec{J}\,) &= -\frac{\partial \rho}{\partial t} \neq 0 \\
\nabla \cdot (\nabla \times \vec{H}) &= \quad 0
\end{aligned} \tag{2.41}$$

To get rid of such an inconsistency, an extra term $\frac{\partial \vec{D}}{\partial t}$ is added to the right term of Eq. (2.34) to yield:

$$\nabla \times \vec{H} = \vec{J} + \frac{\partial \vec{D}}{\partial t} \tag{2.42}$$

Actually, $\frac{\partial \vec{D}}{\partial t}$ represents the displacement current density, since and referring to *Gauss*' law, the electric flux density $\vec{D}$ is directly related to the enclosed electric charge (ρ).

Following the above presented formulation, one can summarize *Maxwell*'s equations as [2]:

$$\nabla \times \vec{E} = -\frac{\partial \vec{B}}{\partial t} + \nabla \times (\vec{u} \times \vec{B}) \tag{2.43}$$

$$\nabla \times \vec{H} = \vec{J} + \frac{\partial \vec{D}}{\partial t} \tag{2.44}$$

$$\nabla \cdot \vec{B} = 0 \tag{2.45}$$

$$\nabla \cdot \vec{D} = \rho \tag{2.46}$$

2.4 Electromagnetic Models

Different electromagnetic models could be derived from the *Maxwell* equations depending on the problem. One can distinguish the two main used ones as illustrated in Fig. 2.12:

- Magnetostatic (Fig. 2.12a): distribution of static magnetic field due to the presence of magnets and currents,
- Magnetodynamic (Fig. 2.12b): distribution of magnetic field and eddy current due to moving magnets and time variable currents.

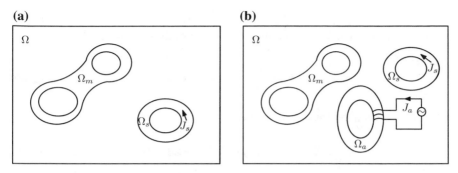

Fig. 2.12 Main components of magnetostatic (**a**) and magnetodynamic (**b**) configurations. **Legend**: Ω studied domain, Ω_m magnetic domain, Ω_a active conductor, Ω_s inductor

2.4.1 Boundary Conditions

The behaviour of $\phi(p, t)$ on the boundary Γ of the studied domain Ω is described as boundary conditions. Among the most used boundary conditions, one can distinguish [4]:

$$
\begin{aligned}
\phi/_{\Gamma_1} &= 0; & \phi/_{\Gamma_1} &= \phi_n & &\text{called } \textit{Dirichet} \text{ condition.}\\
\frac{d\phi/_{\Gamma_2}}{dn} &= 0; & \frac{d\phi/_{\Gamma_2}}{dn} + k\phi &= \phi_n & &\text{called } \textit{Newmann} \text{ condition.}
\end{aligned}
\tag{2.47}
$$

where n is the normal to the boundary Γ which is composed of two part Γ_1 and Γ_2.

2.4.2 Magnetostatic

Magnetostatic focuses on the calculation of the magnetic field whose variation is independent of time. More specifically, magnetostatic model aims to evaluate magnetic fields when the sources of these fields are known. The two possible sources for magnetic fields are (i) coils carrying electric currents and (ii) permanent magnet materials.

The fundamental relations of magnetostatic are deduced from *Maxwell's* equations with considering removing derivatives with respect to time. So, the equations of electricity and magnetism are decoupled, leading to separate study of electrostatic and magnetostatic.

The fundamental relations of magnetostatic, written in their local form, are [4]:

$$
\vec{B} = \mu \cdot \vec{H}; \nabla \cdot \vec{B} = 0; \nabla \times \vec{H} = \vec{J} \left(\frac{\partial \vec{D}}{\partial t} = 0 \right);
\tag{2.48}
$$

We may consider a potential $\overrightarrow{A}$ so that one can express $\overrightarrow{B} = \nabla \times \overrightarrow{A}$. This is possible since the divergence of a rotor field is zero. Consequently:

$$\nabla \times \overrightarrow{H} = \nabla \times \frac{\overrightarrow{B}}{\mu} = \nabla \times (\frac{1}{\mu} \nabla \times \overrightarrow{A}) = \overrightarrow{J} \tag{2.49}$$

which can be expressed as the *Poisson* equation:

$$\nabla \times (\nabla \times \overrightarrow{A}) = \nabla^2 \overrightarrow{A} = \mu \cdot \overrightarrow{J} \tag{2.50}$$

The *Laplacian* of $\overrightarrow{A}$, meaning $\nabla^2 \overrightarrow{A}$ in the global Cartesian coordinate system is:

$$\begin{aligned}\nabla^2 \overrightarrow{A} &= (\frac{\partial^2 A_x}{\partial x^2} + \frac{\partial^2 A_y}{\partial y^2} + \frac{\partial^2 A_z}{\partial z^2}) \overrightarrow{x} + (\frac{\partial^2 A_x}{\partial x^2} + \frac{\partial^2 A_y}{\partial y^2} + \frac{\partial^2 A_z}{\partial z^2}) \overrightarrow{y} \\ &+ (\frac{\partial^2 A_x}{\partial x^2} + \frac{\partial^2 A_y}{\partial y^2} + \frac{\partial^2 A_z}{\partial z^2}) \overrightarrow{z}\end{aligned} \tag{2.51}$$

The above cited equation could be greatly simplified if the current density vector $\overrightarrow{J}$ is collinear with one axis direction ("x" or "y'" or "z").

When the current density $\overrightarrow{J}$ is null in the studied domain, the Eq. (2.50) turns to be:

$$\nabla^2 \overrightarrow{A} = 0 \tag{2.52}$$

This is called the Laplace's equation, and $\overrightarrow{A}$ is called magnetic vector potential.

2.4.3   Magnetodynamics

The fundamental relations of magnetodynamic are deduced from *Maxwell's* equations, where the electric and magnetic fields are coupled. Such a formulation has to take into account the time-varying fields and the possible movement of the considered current carrying coils.

Although simplifying assumptions outlined in the preceding sections yielded reasonable complex problems in the case of simple geometries, electrical actuators present more complicated problem. In an attempt to improve the simplicity of the proposed formulation, let us consider a simple schematic problem of a linear actuator as illustrated in Fig. 2.13.

The analytical formulation is based on some geometrical parameters of the considered actuator. To do so, let us call:

- p is the number of pole pairs,
- q is the number of phases,
- $\omega = 2\pi f$, f is the primary frequency,
- τ is the pole pitch or half the period of the primary magnetomotive force wave.

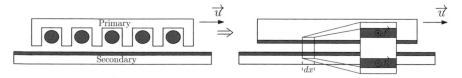

Fig. 2.13 Schematic view of a linear actuator where the primary excitation current is modeled as a fictitious distributed current sheet

The current carried by the primary windings can be replaced by a fictitious and infinitely thin layer of current distributed over the surface of the stator facing the air gap. This current is called the "current sheet" and it produces the same sinusoidal magnetomotive force (mmf) in the air gap as the fundamental component of the one produced by the conductors.

The primary equivalent current sheet is supposed to be excited by the current $\vec{J_1}$:

$$\vec{J_1} = J_m \exp\left(j(\omega t - kx)\right) \tag{2.53}$$

where k is a constant that equals π/τ and J_m is the amplitude of the primary equivalent current sheet, with [4]:

$$J_m = \frac{\sqrt{2}qk_wN_{ph}I_1}{p\tau} \tag{2.54}$$

where N_{ph} is the number of turns per phase in series, I_1 is the rms value of the primary current, k_w is the winding factor of the fundamental.

The proposed magnetodynamic formulation is aimed at the determination of the vector potential $\vec{A}$, considering two case studies: (i) one-dimensional physical model and (ii) two-dimensional physical model.

The current study is based on the fundamental relations of magnetodynamic. Theses are written in their local form, such as [6]:

$$\nabla \times \vec{H} = \vec{J} \tag{2.55}$$

$$\nabla \times \vec{E} = -\frac{\partial \vec{B}}{\partial t} \tag{2.56}$$

$$\nabla \cdot \vec{B} = 0 \tag{2.57}$$

$$\vec{B} = \mu\vec{H} \tag{2.58}$$

$$\vec{J} = \sigma(\vec{E} + \vec{u} \times \vec{B}) \tag{2.59}$$

where $\vec{H}$, $\vec{J}$, $\vec{B}$, $\vec{E}$ and $\vec{u}$ are vectors for the magnetic field intensity, current density, magnetic flux density, electrical field intensity, and primary moving velocity, respectively. μ is the permeability and σ is the conductivity of the considered material.

2.4.3.1   1-D Physical Model

By the introduction of vector potential $\vec{A}$, we can get other expressions for $\vec{B}$ and $\vec{E}$:

$$\vec{B} = \nabla \times \vec{A} \tag{2.60}$$

$$\vec{E} = -\frac{\partial \vec{A}}{\partial t} \tag{2.61}$$

In the manner of the *Ampere*'s law ($\nabla \times \vec{B} = \mu_0 \vec{J}$) applied to the closed loop of Fig. 2.13 and under the assumption of the infinite permeability of the iron, Eq. (2.55) turns to be:

$$\frac{g_t}{\mu_0} \frac{\partial \vec{B}_y}{\partial x} = \vec{J}_1 + \vec{J}_2 \tag{2.62}$$

where $\vec{J}_1$ and $\vec{J}_2$ are the complex form of the equivalent current sheet in the primary and the secondary, respectively. $\vec{B}_y$ is the y-axis component of the flux density in the air gap region.

Referring to the actuator concept, the active part of the primary current $\vec{J}_1$ is limited to its z component. So, the vector potential deduced from Eq. (2.62) has also only a z-axis component and Eqs. (2.60) and (2.61) can be further expressed by:

$$\vec{B}_y = -\frac{\partial \vec{A}_z}{\partial x} \tag{2.63}$$

$$\vec{E}_z = -\frac{\partial \vec{A}_z}{\partial t} \tag{2.64}$$

Combining Eqs. (2.63), (2.64), and (2.59), one can write:

$$\vec{J}_2 = -\sigma_e \left(\frac{\partial \vec{A}_z}{\partial t} + u \frac{\partial \vec{A}_z}{\partial x} \right) \tag{2.65}$$

where σ_e is the surface conductivity in the amount of $g_t \sigma$, with g_t is the total magnetic air gap length and $u = u_x$ is the primary moving velocity along the x-axis.

Moreover, $\vec{A}_z$ is function of time t and position x and it could be expressed as:

$$\vec{A}_z = A_z(x, t) = A_z \exp^{j(\omega t - kx)} \tag{2.66}$$

Based on the above formulation, Eq. (2.62) turns to be:

$$\frac{g_t}{\mu_0} \frac{\partial^2 \vec{A}_z}{\partial x^2} - \sigma_e u \frac{\partial \vec{A}_z}{\partial x} - j\omega \sigma_e \vec{A}_z = -J_m \exp^{j(\omega t - kx)} \tag{2.67}$$

which is the *Helmholtz* equation and whose solution could be found out as [6]:

$$\overrightarrow{A}_z = c_s \exp^{j(\omega t - kx)} + cc_1 \exp^{(-\frac{x}{\alpha_1} + j(\omega t - \frac{\pi}{\tau_e}x))} + cc_2 \exp^{(\frac{x}{\alpha_2} + j(\omega t + \frac{\pi}{\tau_e}x))} \tag{2.68}$$

where c_s, cc_1, and cc_2 are coefficients decided by boundary conditions, α_1 and α_2 are attenuating coefficients of the air-gap entrance- and exit-flux density waves, respectively, and τ_e is the end-effect half-wave length. The different expressions of theses coefficients are listed in [6].

2.4.3.2 2-D Physical Model

Referring to Eq. (2.55), and in the case of two dimensional physical model, the application of the *Ampere*'s law to the rectangle in Fig. 2.13 leads to [4]:

$$\frac{g_t}{\mu_0} \left(\frac{\partial^2 \overrightarrow{A}_z}{\partial x^2} + \frac{\partial^2 \overrightarrow{A}_z}{\partial y^2} \right) - \sigma_e u \frac{\partial \overrightarrow{A}_z}{\partial x} - j\omega \sigma_e \overrightarrow{A}_z = -J_m \exp^{j(\omega t - kx)} \tag{2.69}$$

Equation (2.69) written in the secondary coordinate system, where s is the slip, turns to be:

$$\frac{g_t}{\mu_0} \left(\frac{\partial^2 \overrightarrow{A}_z}{\partial x^2} + \frac{\partial^2 \overrightarrow{A}_z}{\partial y^2} \right) - js\omega \sigma_e \overrightarrow{A}_z = -J_m \exp^{j(s\omega t - kx)} \tag{2.70}$$

The solution of such an equation could be expressed as a function of (x, y, t) variation such as:

$$\overrightarrow{A}_z(x, y, t) = A_z(y) \exp^{j(s\omega t - kx)} \tag{2.71}$$

where $A_z(y)$ depends on the normal position within the air gap.

The resolution of the Eq. (2.70) could be achieved using a separation of variables. $\overrightarrow{A}_{3z}(x, y, t)$ is thus [4]:

$$\overrightarrow{A}_{3z}(x, y, t) = (c_1 \exp^{\gamma y} + c_2 \exp^{-\gamma y}) \exp^{j(s\omega t - kx)} \tag{2.72}$$

where γ is the solution of the corresponding characteristic equation, and c_1 and c_2 are two constant which could be determined based on the boundary condition.

2.5 Energy Relations

Electromechanical energy conversion occurs when coupling fields are disturbed in such a way that the energy stored in the fields changes with mechanical motion. A justification of this statement is possible from energy conservation principles, which will enable us to determine the magnitudes of mechanical forces arising from magnetic field effects.

Considering the case of real energy conversion system, one can define the total electric energy provided by the electric source W_{se} as the sum of [7]:

- the energy dissipated as *Joule* losses W_{je},
- the energy corresponding to the increase in stored magnetic energy W_m,
- the electric energy corresponding to the achieved mechanical work W_{me}.

where the equivalent mechanical energy W_{me}, is partially available as useful output mechanical energy W_{ume}.

Actually, referring to Fig. 2.14, the mechanical energy W_{me} expression could be defined as:

$$W_{me} = W_{ume} + W_{lme} + W_{sme} \tag{2.73}$$

where W_{lme} and W_{sme} are the mechanical losses and the stored mechanical energies, respectively.

Although, the *Joule* losses are inevitably present, they have no influence in the energy conversion process [7]. Moreover, this latter is not affected by the repartition of the mechanical energy on useful part and losses.

Based on the above cited statements, only the conservative portion of a system is considered. Indeed, assuming a conservative system and neglecting the *Joule* losses, one can write the total electric energy provided by the electric source W_{se} as the sum of:

- the energy corresponding to the increase in stored magnetic energy W_m,
- the electric energy corresponding to the achieved mechanical work W_{me}.

$$W_{se} = W_{me} + W_m \tag{2.74}$$

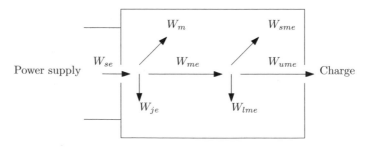

Fig. 2.14 Real energy conversion system [7]

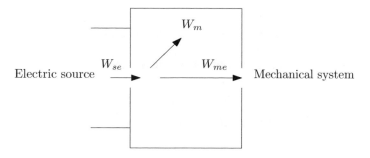

Fig. 2.15 Conservative energy conversion system with neglecting the *Joule* losses

Equation (2.74) constitutes the basics of the electromechanical converter operating formulation. This is schematically represented in Fig. 2.15 [7].

Based on the expressions of the small variation of the involved energies, Eq. (2.74) yields:

$$\mathcal{V}idt = dW_{me} + dW_m \tag{2.75}$$

where $\mathcal{V}$ and i denotes the voltage and current of the electric source and dW_m is the increase in stored magnetic energy.

Referring to Eqs. (2.23) and (2.75) turns to be:

$$dW_{me} = -dW_m + id\phi \tag{2.76}$$

with $d\phi = nd\varphi$, is the incremental variation of the flux linkage.

The achieved mechanical work dW_{me} is linked to the force of electrical origin F_{em}. The general vectorial expression of this latter is:

$$\overrightarrow{F}_{em} = \frac{\partial W_{me}}{\partial x}\overrightarrow{x} + \frac{\partial W_{me}}{\partial y}\overrightarrow{y} + \frac{\partial W_{me}}{\partial z}\overrightarrow{z} \tag{2.77}$$

The study is focused towards the investigation of an electromagnetic system where the motion is supposed collinear with one axis direction (x-axis) and with a simple current excitation (i). This leads to:

$$dW_{me} = F_{em}dx \tag{2.78}$$

Prior to the determination of the electromagnetic force, the following section is devoted to the formulation of the magnetic energy concept. Indeed and referring to Eq. (2.76), the electromagnetic force is deduced form the assessment of the magnetic phenomenon. To do so and in an attempt to simplify the proposed formulation, the following assumptions are adopted:

- the flux leakage is ignored,
- the excitation winding is supposed to be kept unchanged (no movement or deformation),
- the eddy current and the hysteresis effects are ignored.

2.5.1   Magnetic Energy and Co-energy

In order to evaluate the magnetic energy, let us consider the system in absence of motion. Under these conditions and based on Eq. (2.76), the magnetic stored energy is equal to the input electric energy and it could be expressed as:

$$dW_m = \mathcal{V}idt = id\phi = \xi d\varphi \qquad (2.79)$$

Referring to Eq. (2.79), it appears that any flux variation is associated to a variation of the electrical energy applied to the conservative system.

Such a flux variation is due to:

- a variation of the source current or voltage,
- a modification of the magnetic circuit following a movement of some pieces,
- the two above cited phenomenon simultaneously.

In absence of movement, the current is only function of the flux which is represented by the $\phi = f(i)$ (or $B = f(H)$) curve. Such a function depends on:

- the number of turn of the excitation coil,
- the material proprieties and geometries of the magnetic circuit,
- the chosen position of the moving part.

Let us consider an electromagnetic system characterized by a linear behaviour. The characteristic $\phi = f(i)$ for a given position of the mover is a straight line as shown in Fig. 2.16. It is to be noted that the electromagnetic system is assumed linear if a large part of the flux line path is located within iron.

Basically and referring to Eq. (2.22), the flux density B could be expressed based on flux ϕ and the equivalent flux tube surface S, as:

$$B = \frac{\phi}{S} \qquad (2.80)$$

Considering the *Ampere* law, the elementary magnetic energy, represented by the hachured elementary area in Fig. 2.16, turns to be:

$$dW_m = id\phi = HlSdB \qquad (2.81)$$

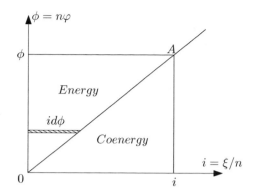

Fig. 2.16 Flux-current characteristic in the case of linear system

Considering the linear case, the integration of Eq. (2.81) leads the determination of the magnetic energy density W_{md}:

$$W_m = \int_0^B HV\,dB \quad \Rightarrow \quad W_{md} = \frac{B^2}{2\mu}V \tag{2.82}$$

where $V = lS$ is the equivalent volume of the considered system.

Referring to Fig. 2.16, the linear characteristic could be expressed considering the self inductance L which is independent on the current i, leading to the following relation [3]:

$$\phi = Li \tag{2.83}$$

Actually and based on electrical *Ohm* law by one hand and the magnetic *Ohm* law (Eq. 2.38), by the other hand, one can express the self inductance L, such as:

$$L = \frac{n^2}{\mathcal{R}} \tag{2.84}$$

Based on Eq. (2.83), the integration of Eq. (2.79) leads to the determination of the magnetic energy density W_m:

$$W_m = \int_0^\phi i\,d\phi = \frac{1}{2}Li^2 \tag{2.85}$$

Based on its definition, the magnetic energy W_m corresponds to the energy stocked in the magnetic circuit. It consists in the provided amount of energy leading to the increase of the magnetic field in the considered circuit from zero to ϕ. Indeed, the magnetic energy W_m corresponds to the surface of the upper triangle as shown in Fig. 2.16. However, the co-energy W_m' is not related to a physical phenomenon and it is defined as follows:

$$W_m + W_m' = i\phi = \xi\varphi \tag{2.86}$$

Fig. 2.17 Flux-current
characteristic in the case of
non linear system

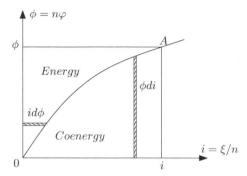

The magnetic co-energy W'_m corresponds to the surface of the lower triangle as illustrated in Fig. 2.16, and it is expressed as:

$$W'_m = \int_0^i \phi di = \int_0^\xi \varphi d\xi \qquad (2.87)$$

It is to be noted that for the case of linear electromagnetic system the two quantities W_m and W'_m are equal.

2.5.2 Electromagnetic Force Formulation

The investigation of the electromagnetic force is considered taking into consideration the saturation effect of the involved materials. In that case, the magnetic energy W_m and co-energy W'_m which are state functions of the conservative system are no more equal and the electromagnetic system is no longer linear.

Indeed, with considering the saturation effects, the electromagnetic system is characterized by a non linear function ($\phi = \tilde{f}(i)$) as illustrated in Fig. 2.17.

In what follows, the evaluation of the electromagnetic force F_{em} goes through the determination of the achieved mechanical work dW_{me} following a given displacement dx.

Such a force is the result of the interaction between moving and fixed parts of the considered electromagnetic system. To do so, the three variables the mechanical one x, the magnetic one ϕ and the electrical one i, should be taken into consideration.

Two particular cases are considered for which two among the three considered variables could be considered independents. In fact, under particular conditions either (i, x) or (ϕ, x) may be considered as independent variables. The choice is mainly depending on the speed of the considered displacement dx:

- Considering a very fast displacement of the moving part from position x to $x + \Delta x$, one can assume that the flux ϕ is considered constant [7]. Indeed, the change of the current from an initial value i_1 to final one i_2 could be done instantly. While, a very fast variation of the flux ϕ yields that the quantity $\frac{d\phi}{dt}$ has to be infinite, which is not possible. Under these condition (i, x) could be considered as independent variables,

• Considering a very slow displacement of the moving part from position x to position $x + \Delta x$, one can assume that the current i is kept constant as the quantity $\frac{d\phi}{dt}$ is almost null [7]. Under these conditions (ϕ, x) could be considered as independent variables.

In what follows, and to evaluate the electromagnetic force in whatever condition, let us consider the evolution of the electromagnetic system under small changes.

Let us consider first (i, x) as independent variable. So, the flux linkage ϕ is given by $\phi = \phi(i, x)$, which can be expressed in terms of small changes as:

$$d\phi = \frac{\partial \phi}{\partial i} di + \frac{\partial \phi}{\partial x} dx \qquad (2.88)$$

In the same manner, $W_m(i, x)$, so that:

$$dW_m = \frac{\partial W_m}{\partial i} di + \frac{\partial W_m}{\partial x} dx \qquad (2.89)$$

Thus, Eqs. (2.88) and (2.89), when substituted into Eq. (2.90), yield:

$$F_{em} dx = \left(\frac{-\partial W_m}{\partial x} + i \frac{\partial \phi}{\partial x} \right) dx + \left(\frac{-\partial W_m}{\partial i} + i \frac{\partial \phi}{\partial i} \right) di \qquad (2.90)$$

Because the incremental changes di and dx are arbitrary, F_{em} must be independent of these changes. Thus, to make F_{em} independent of di, its coefficient in Eq. (2.90) must be equal to zero. Consequently, Eq. (2.90) becomes:

$$F_{em} = \frac{-\partial W_m}{\partial x}(i, x) + i \frac{\partial \phi}{\partial x}(i, x) \qquad (2.91)$$

Since the flux ϕ is considered constant under the assumption of very fast movement, Eq. (2.91) is limited to its first term:

$$F_{em} = \frac{-\partial W_m}{\partial x}(i, x) \qquad (2.92)$$

which is the electromagnetic force equation and it holds true if i is the independent variable.

Referring to Eq. (2.86), one can write the variational form as:

$$dW_m + dW'_m = i d\phi + \phi di \qquad (2.93)$$

Based on Eqs. (2.76), (2.78) and (2.93), one can establish that:

$$dW_m = id\phi - F_{em}dx$$
$$dW'_m = \phi di + F_{em}dx$$

(2.94)

Considering the co-energy dW_m' expression and accounting for the fact that ϕ is supposed constant, the electromagnetic force F_{em} is expressed as:

$$F_{em} = \frac{\partial W_m'}{\partial x}(i, x)$$

(2.95)

Equation (2.95) could be expended at its general form as:

$$F_{em} = \frac{\partial W_m'}{\partial x}\vec{x} + \frac{\partial W_m'}{\partial y}\vec{y} + \frac{\partial W_m'}{\partial z}\vec{z}$$

(2.96)

when ϕ is taken as the independent variable, one can write $i = i(\phi, x)$ and $W_m = W_m(\phi, x)$. Considering the above relations, the variational form of the magnetic energy is expressed as:

$$dW_m = \frac{\partial W_m}{\partial \phi}d\phi + \frac{\partial W_m}{\partial x}dx$$

(2.97)

whose substitution in Eq. (2.76), gives:

$$F_{em}dx = -\frac{\partial W_m}{\partial x}dx - \frac{\partial W_m}{\partial \phi}d\phi + id\phi$$

(2.98)

Taking into account that $\frac{\partial W_m}{\partial \phi} = i$, Eq. (2.98) finally becomes:

$$F_{em} = -\frac{\partial W_m}{\partial x}(\phi, x)$$

(2.99)

Equation (2.99) could be expanded at its general form as:

$$F_{em} = -\frac{\partial W_m}{\partial x}\vec{x} - \frac{\partial W_m}{\partial y}\vec{y} - \frac{\partial W_m}{\partial z}\vec{z}$$

(2.100)

2.6 Conclusion

This chapter was aimed at an analytical modelling of the electromagnetic forces exhibited by linear actuators. The study was initiated by an overview of electromagnetic phenomenon. To do so, electric and magnetic material properties and specifications were presented. A formulation of the air gap flux density trough the resolution of governing equation on the magnetic vector potential was then carried out, considering two cases:

- magnetostatic model characterised by a static magnetic field distribution due to the presence of magnets and currents,
- magnetodynamic model characterised by a time dependent distribution of magnetic field and eddy current due to moving magnets and time variable currents.

The established models were based on the main electromagnetic laws and the Maxwell equations. Moreover, a special attention has been paid to the energy relations. Indeed, considering a conservative electromagnetic system, the energy and the co-energy relations were developed dealing with linear and non linear systems. Such a formulations where treated in an attempt to predict the developed electromagnetic force.

Increasing the electromagnetic force production capability represents a crucial energy efficiency benefit which is targeted in several applications such as wave energy harvesting and MAGLEV trains. These will be treated in depth in Chaps. 3 and 4, respectively.

References

1. B. Thidé, *Electromagnetic Field Theory* (Upsilon Books, Uppsala, Sweden, 2004)
2. H.E. Knoepfel, *Magnetic Fields: A Comprehensive Theoretical Treatise For Practical Use* (Wiley, New York, USA, 1999)
3. M. Jufer, *Electro-Mechanics, Electrics, Electronics and Electrical Engineering (in frensh)* (Dumas press, Dunod, France, 1983)
4. I. Boldea, *Linear Electric Machines, Drives, and MAGLEVS Handbook*, (CRC Press, Taylor and Francis Group, New York, 2013), pp. 331–368
5. I. Boldea, S.A. Nasar, *Linear Electric Actuators and Generators* (Cambridge University Press, London, UK, 1997)
6. W. Xu, J.G. Zhu, Y. Zhang, Y. Li, Y. Wang, Y. Guo, An improved equivalent circuit mdel of a single-sided linear induction motor. IEEE Trans. Vehicular Technol. **59**(5), 2277–2289 (2010)
7. J. Lesenne, F. Notelet, G. Séguier, *Introduction to the Deep Electrotechnic (in frensh)* (Technique and Documentation, Paris, France, 1981)

Chapter 3
Tubular-Linear Synchronous Machines: Application to Wave Energy Harvesting

Abstract The chapter is aimed at one of the emergent sustainable applications, that is wave energy (WE) harvesting using appropriate converters (WECs). The survey is started by a classification of WECs according to the technology of their power take-off systems with emphasis on the topology of the integrated generator including rotating and linear topologies. Linear PM synchronous machines represent the most viable candidates thanks to their high force density and energy efficiency at low speeds. Of particular interest is the inset PM (IPM) tubular topology which offers an increase of the energy efficiency and an intrinsic-cancellation of the radial attractive forces. The IPM tubular-linear synchronous machines (T-LSMs) are then the subject of modelling considering its magnetic equivalent circuit (MEC). Following its synthesis, the MEC is solved considering an iterative numerical procedure. The preliminary results reveal that the IPM T-LSM features are affected by the end effect. Two design approaches dedicated to the minimization of such a drawback are proposed and their effectiveness checked by finite element analysis. The survey is achieved by an extension of the validity of the proposed model to the time-varying features by incorporating the mover position in the MEC.

Keywords Wave energy converters
Inset PM tubular-linear synchronous machines · Magnetic equivalent circuit
End effect minimization · Mover position incorporation · Finite element analysis

3.1 Introduction

Following the 70th oil crisis, the world realized for the first time what it would be like if fuels would no longer be cheap or unavailable. Facing such situation, renewable energies have been the subject of an intensive regain of interest. Many research and developed projects were launched so far, with emphasis on the investigation of the power potential of classical and emergent earth's natural energy reserves. A particular interest has been and continue to be paid to the wind and solar energies which are presently considered as viable candidates to assist, in a short term, then substitute, in a long term, the carbon-based energy.

© Springer Nature Singapore Pte Ltd. 2019
A. Souissi et al., *Linear Synchronous Machines*, Power Systems,
https://doi.org/10.1007/978-981-13-0423-1_3

Beyond these conventional renewable energies, an increasing attention is presently given to emergent green sources exhibiting promising potentials, such as, the geothermal, biomass and ocean energies. Dealing with the latter, it should be underlined that the oceans cover roughly 75% of the earth's surface which is a synonym of a great energy potential. Indeed, the oceans are continuously producing huge amounts of energy in several forms, including:

- tidal,
- marine current,
- temperature-gradient,
- salinity,
- wave.

Fundamentally, waves exhibit an energy density higher than all other forms of marine energy and even than the wind one, enabling dedicated conversion systems to harvest higher power. Moreover, wave energy is more available and more predictable with better demand matching. In spite of the huge potential of wave energy, its conversion into electricity is currently far from being a mature technology which makes its current market penetration insignificant. In order to face this challenging technological gap, different concepts dedicated to wave energy harvesting are being developed and investigated worldwide.

Since 2002, over 1000 wave energy converters (WECs) have been patented in Japan, Europe, and North America [1]. Giving the diversity of WECs, they have been categorized according to their:

◇ shore location,
◇ principle of operation,
◇ generator topology.

as illustrated in Fig. 3.1.

This said, it should be underlined that the shore-located WECs have received limited interest due to the long, costly, and inefficient pipe infrastructure required to transport the sea water from the offshore or near-shore to shore, and the associated significant power loss. Offshore or near-shore WECs could classified according to three families, namely [2]:

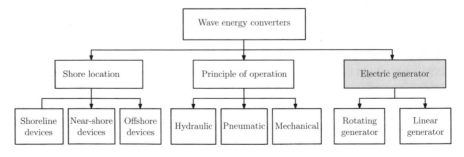

Fig. 3.1 Classification of the wave energy converters according to different criteria

- Attenuators which are floating on the sea surface pointing to a direction parallel to the predominant wave one. They are riding the waves causing their deformation which is converted into electricity by means of appropriate devices. The most popular attenuator is the so-called "Pelamis" which has been developed by *Pelamis Wave Power*.
- Point absorbers, which unlike the attenuators are characterized by a geometry relatively small compared to the incident wavelength. They could be either:
  - submerged below the surface relying on pressure differential. The *Archimedes* wave swing, shown in Fig. 1.15, belongs to this class of point absorbers, or
  - floating, oscillating up and down, on the surface of the water. The different topologies of the heaving buoy illustrated in Fig. 1.16 are of this type of point absorber class.
- Terminators have their principal axis perpendicular to the predominant wave direction in such a way that they are intercepted, instead of riding, waves. Limited number of terminator concepts have been reported in the literature. A typical example of this type of terminator-WE is the one developed at the University of Edinburgh (UK), the so-named "the Salter's Duck".

3.2 Generator Topology-Based Classification of WECs

Figure 3.2 exhibits the diverse technologies of the power take-off (PTO) systems integrated in WECs. Focusing the conversion end chain, it is to be noted that all WECs equipped with rotating machines integrate devices, such as gearboxes, pistons, air turbines, and water turbines, as power take-off systems which are devoted to interface of the wave hydraulic power to the generator shaft. Three major WEC topologies, equipped with rotating generators, have been reported in the literature which are: (i) the pelamis [3], (ii) the wave Dragon [4], and (iii) the oscillating water column [5].

3.2.1 Rotating Generators

In the manner of wind energy harvesting, the generators used in WECs need to cope with the variable speed character of the electromechanical conversion resulting from the intermittency of the wave hydraulic power. Within this statement, WECs could be suitably-equipped with two possible viable candidates which are: (i) the doubly fed induction generator (DFIG) and (ii) the permanent magnet synchronous machine (PMSG). These are discussed hereunder.

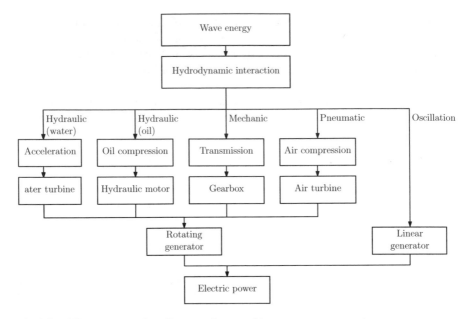

Fig. 3.2 Different power take-off systems integrated in wave energy converters

3.2.1.1 Doubly-Fed Induction Generators

The DFIG is topologically-identical to the wound rotor induction machine. It is equipped with three balanced wound phases inserted in slots located on both stator and rotor air gap sides. The rotor phases are electrically-accessed through a slip-ring system. However, the two machines differ by their principle of operation. While the one of the induction machine relays on the induction phenomenon, the principle of operation of the DFIG is based on the synchronization of the stator and rotor rotating fields according to two following scenarios:

- the stator and rotor rotating fields are turning in the same direction at speeds Ω_s and Ω_r, respectively. In this case, their synchronization is achieved by a mechanical speed Ω_m lower than the one of the resultant rotating field, yielding the so-called "hyposynchronism" which is characterized by a slip s_{hypo} expressed as:

$$s_{hypo} = \frac{\Omega_s - \Omega_m}{\Omega_s} = \frac{\Omega_r}{\Omega_s} > 0 \tag{3.1}$$

- the stator and rotor rotating fields are turning in opposite directions. In this case, their synchronization is achieved by a mechanical speed higher than the one of the resultant rotating field, yielding the so-called "hypersynchronism" with a slip s_{hyper} expressed as:

$$s_{hypo} = \frac{\Omega_s - \Omega_m}{\Omega_s} = -\frac{\Omega_r}{\Omega_s} < 0 \qquad (3.2)$$

With this said, a particular scenario, characterized by a rotation in the same direction and at the same speed of the stator rotating field and the machine shaft (a slip equal to zero), should be underlined. As a result, the rotor angular frequency turns to be null (DC current). In other words, the synchronous machine operation corresponds to a particular case of the DFIG one.

From a power flow point of view, the DFIG enables the conversion of the mechanical power P_m applied to the shaft into an electrical power P_s directly delivered by the stator circuits to the load (case of stand alone operation) or to the grid. The rotor circuits are fed by the stator ones through two cascaded AC-DC converters sharing the same DC bus. As far as the rotor power P_r is limited by the slip range, not exceeding $\pm30\%$, the rating of the static converter in rotorcircuits represents a fraction of the DFIG one. This represents a crucial cost benefit.

However, DFIGs suffer from the required periodical maintenance of the slip-ring system, with a relatively higher frequency than in wind generating systems, due to the greater corrosion effect. Moreover, due to their low pole pair number and for the sake of their suitable interface to the grid, a speed multiplier has to be integrated in between their shaft and prime mover. Beyond its fallout on the system cost-effectiveness and compactness, the speed multiplier also requires a periodical maintenance.

3.2.1.2 Permanent Magnet Synchronous Generators

PM excited synchronous machines have been and continue to be given an increasing attention in variable speed generating systems. This trend is motivated by the following advantages:

- a high torque to volume ratio. This enables the improvement of the compactness. Such an improvement is reinforced by the elimination of the brush-ring system which is required in the case of field-excited synchronous machines,
- a high efficiency gained by the elimination of the rotor *Joule* losses. Furthermore, in the case of high remanence PMs and for a given loading level, the armature current is lower than in field-excited machines, which leads to lower armature *Joule* losses,
- maintenance free.

In spite of their advantages, PM synchronous machines suffer from some shortcomings, such as:

- a limited flux weakening range. This limitation could be overcome by substituting the distributed winding in the armature by fractional-slot concentrated ones. However, this rearrangement of the armature is penalized by an excessive increase of the eddy current losses in the PMs which is due to the dense harmonic content caused by the armature magnetic reaction. A reduction of these losses could be achieved by the PM segmentation even thought this may compromise the machine manufacturing and cost-effectiveness,

- a high cost compared to field-excited machines in the case of rare earth PMs which are the most attractive ones given their high performance,
- the risk of demagnetization caused by an excessive armature magnetic reaction or by another excitation source creating a flux circulating in an opposite direction to the one of the PM magnetization.

PM synchronous generators (PMSGs) could be viable candidates for wave energy harvesting. They make it possible the eradication of the maintenance problem penalizing DFIGs. Indeed, beyond the elimination of the brush-ring system, they are suitably-adapted to designs with high pole pair numbers which enables the elimination of the speed multiplier. Nevertheless, the cost-effectiveness of the wave energy harvesting systems is significantly affected using PMSGs. Indeed, beyond the alarming PM cost increase, the harvested power is transferred from the armature to the grid through two cascaded AC-DC converters with a rating similar to the PMSG one which represents a crucial cost loss.

3.2.2 Linear Generators

WECs equipped with rotating generators systematically-require PTO mechanisms that interface the wave oscillating motion to the machine shaft. Despite their optimized design, these devices affect the energy efficiency of the WECs as well as their reliability and cost-effectiveness. Based on this statement, the substitution of the rotating generators by linear ones, yielding the so-called "direct drive" WECs, represents an alternative to face the limitations allied to the PTO mechanisms. With this said, and beyond the high power density, efficiency, and reliability exhibited by PM synchronous generators, some specificities related to wave energy should be underlined, such that:

- giving the fact that the wave oscillating speed in varying from 0.5 to 2 m/s, the generator high efficiency range has to be extended to low speed operation,
- the generator should exhibit a high peak force. For instance, for a generator designed to produce a maximum power of 100 kW, a maximum force of 100 kN is required at a velocity of 1 m/s,
- the generator should develop a high average power despite the irregular oscillation motion characterizing the wave motion. This could be achieved by the selection of suitable power electronic converters interfacing the harvested energy to the grid.

Other critical design issues, allied to mechanical constraints, have to be taken into consideration, especially [6]:

⋄ the high attractive forces between the translator and the stator particularly in flat linear generator topologies,
⋄ the relatively large air gap mandated by the mechanical manufacturing tolerances, the limited stiffness, and the corrosion.

Several topologies of linear generators could be integrated in WECs, among which one can distinguish:

✓ the switched reluctance machines,
✓ the PM synchronous machines.

Linear Induction generators are rarely incorporated in direct drive WEC systems. This is due to the fact they are basically-designed to operate close to the synchronous speed which is not in harmony with the wave energy intermittency.

3.2.2.1 Switched Reluctance Generators

Linear switched reluctance generators (LSRG) have been selected by several research teams [7]. Different topologies have been considered for WEC applications, covering flat double sided [8, 9] and tubular [10–12] concepts. LSRGs are reputed by their simple construction and high mechanical robustness on one hand and by their high reliability and attractive cost-effectiveness on the other hand. Nevertheless, in the manner of rotating switched reluctance machines, linear ones suffer from their low force production capability due to their high leakage inductances [12].

3.2.2.2 Permanent Magnet Synchronous Generators

Linear PM synchronous machines (LPMSMs) are commonly integrated in direct drive WECs, thanks to their high force density and efficiency at low speeds. In addition, a good weight to power ratio could be achieved by increasing the active area of the machine. Referring to Sect. 1.2.2, LPMSMs have three common topologies according to the PMs arrangement in the mover, such that: (i) inset PM machines with an axial magnetization of the magnets [13], (ii) surface-mounted PM machines with a radial magnetization of the magnets [14], and (iii) *Halbach* PM machines with a combined axial-radial magnetization of the magnets [15].

Basically, LPMSMs with radially-magnetized PMs suffer from the following limitations:

• the difficulty to maintain the PMs at the surface of the mover,
• the high cost of the rare earth PMs that need to be used in an attempt to achieve high power densities especially with the additional air gap due to the PM width.

The first limitation is intrinsically-discarded in inset and *Halbach* PM topologies in so far as the magnets are buried in the translator. Furthermore, the second limitation is overcome thanks to the substitution of the the rare earth PMs by ferrite ones. However, the volume of magnets is higher in the *Halbach* PM topology is higher that the one of the inset PM topology which makes this latter more attractive.

With this said, a literature review revealed a variety of LPMSMs with inset PMs (IPMs). They could be designed according to flat geometries with single [16], double [17] or four sides [18], as well as tubular ones [19–21]. This latter is the most suitable topology for wave energy harvesting, thanks to the further increase of the energy efficiency gained with the elimination of the winding end turns so that the armature copper turns to be totally active. Furthermore, the radial attractive forces are diametrically-cancelled which limits the risk of eccentricity and reduces noise and vibration. In what follows, a special attention is paid to the magnetic equivalent circuit (MEC) modelling of IPM tubular-linear synchronous machines (T-LSMs).

3.3   MEC Modelling of IPM T-LSMs

Dealing with electric machine modelling, it is commonly-admitted that the finite element analysis (FEA) approach leads to the most accurate results. However, it requires a high CPU-time which makes it unsuitable for sizing procedures. The MEC-based models represent an interesting tradeoff accuracy/CPU-time. The derivation of the MEC goes through the identification of the different flux tubes, including the main and leakage ones, characterizing the mover-stator flux-linkage. The prediction of the mean length and equivalent surface of each flux tube enables the formulation of its reluctance in terms of the machine geometrical and magnetic parameters considering the *Hopkinson* law, as presented in Chap. 2.

3.3.1   MEC General Resolution Procedure

Basically, if a MEC includes N_b branches and N_n nodes, the number of independent loops is equal $N_l = N_b - N_n + 1$. Let us call $\mathcal{R}$ and $\mathcal{L}$ the reluctance diagonal matrix and the loop MMF vector, respectively. Then, the MMF-flux law can be written as follows:

$$F = \mathcal{R}\Phi \tag{3.3}$$

where F and Φ are the branch MMF and flux vectors, respectively.

It should be noted that the loop and branch MMF vectors are linked by the topological matrix S, as follows:

$$\mathcal{L} = SF \tag{3.4}$$

The topological matrix S includes N_l lines and N_b columns, where the element S_{ij} takes the following values:

$$\begin{cases} \text{``0''} & \text{in the case where the flux of branch } j \text{ is not included in loop } i, \\[1em] \text{``1''} & \text{in the case where the flux of branch } j \text{ is included in loop } i \\ & \text{and flows in a direction similar to the loop orientation,} \\[1em] \text{``}-1\text{''} & \text{in the case where the flux of branch } j \text{ is included in loop } i \\ & \text{and flows in a direction opposite to the loop orientation.} \end{cases}$$

Combining Eqs. (3.3) and (3.4) yields:

$$\mathcal{L} = S\mathcal{R}\Phi \tag{3.5}$$

The node law leads to:

$$\Phi = S^T\psi \tag{3.6}$$

where ψ is the vector of loop fluxes.

Reaching this step, one has to distinguish the two following cases:

- *Case of linear magnetic circuits*: the loop fluxes are calculated considering Eqs. (3.5) and (3.6), as follows:

$$\psi = \left(S\mathcal{R}S^T\right)^{-1}\mathcal{L} \tag{3.7}$$

- *Case of saturated magnetic circuits*: the matrix $\left(S\mathcal{R}S^T\right)$ is no longer invertible. In order to achieve the resolution of the MEC, an error vector $\mathcal{E}$ is defined as follows:

$$\mathcal{E} = \mathcal{F} - \left(S\mathcal{R}S^T\right)\psi \tag{3.8}$$

Considering Eqs. (3.5) and (3.8), the vector of loop fluxes φ is found out using the *Newton-Raphson* algorithm that minimizes $\mathcal{E}$.

A flowchart of the procedure enabling the resolution of the MEC is shown in Fig. 3.3.

3.3.2   Application to IPM T-LSMs

Let us consider the IPM T-LSM concept shown in Fig. 3.4. For a given mover position which is in most if not all cases corresponds to the maximum flux linkage between an armature phase and the mover poles, one can establish a MEC model of the machine taking into consideration the different material (air, iron, PMs, shaft), as illustrated in Fig. 3.4. It highlights the different regions of the machine, such that:

- Region I: mover part,
- Region II: air gap,
- Region III: stator part.

Accounting for the axe-symmetry of the tubular concept, just the half of the machine cross section has been the subject of the MEC modelling.

Let us consider the IPM T-LSM given in Fig. 1.2. The allocation of the stator coils to the armature phases has been carried out using the star of slots approach, as shown in Fig. 3.5.

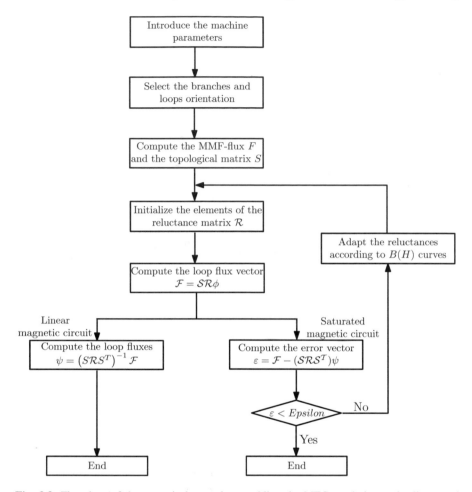

Fig. 3.3 Flowchart of the numerical procedure enabling the MEC resolution under linear and saturated magnetic circuits

Considering the mover position corresponding to the maximum flux linkage with the armature c-phase, the resulting MEC model is illustrated in Fig. 3.6, where the different branches (red color) and loops (green color) are identified. It is to be noted that the orientations of the branches as well as those of the loops have been selected arbitrarily. The MEC includes $N_b = 24$ branches and $N_n = 15$ nodes, which leads to $N_l = 10$ independent loops.

Accounting for the orientations shown in Fig. 3.6, the topological matrix S has been established, considering the methodology presented in Sect. 3.3.1, with 10 rows and 24 columns, as follows:

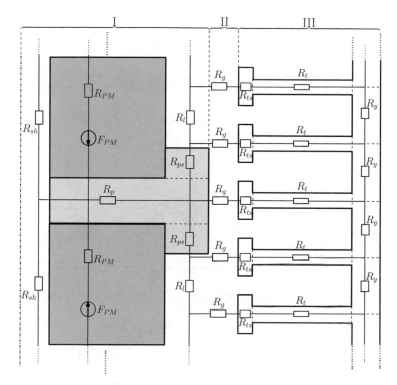

Fig. 3.4 MEC model of an IPM T-LSM concept

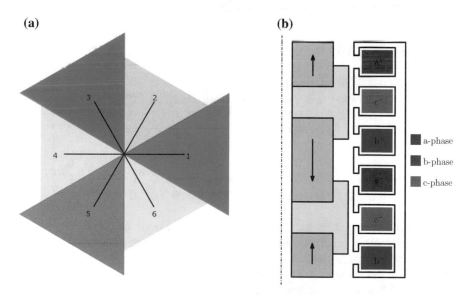

Fig. 3.5 Star of slots-based arrangement of the armature phases of the studied IPM T-LSM. **Legend**: **a** star of slots phasor diagram, **b** stator coil repartition in three armature phases

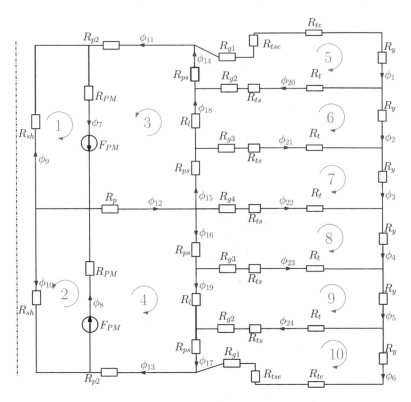

Fig. 3.6 MEC model of the IPM T-LSM under study

$$S = \begin{pmatrix} 0\,0\,0\,0\,0\,0\,1\,0\,1\,0\,0\,0\,0\,0\,0 & 0 & 0\,0\,0 & 0 & 0 & 0 & 0 & 0 \\ 0\,0\,0\,0\,0\,0\,0\,1\,0\,1\,0\,0\,0\,0\,0 & 0 & 0\,0\,0 & 0 & 0 & 0 & 0 & 0 \\ 0\,0\,0\,0\,0\,0\,1\,0\,0\,0\,1\,1\,0\,1\,1 & 0 & 0\,1\,0 & 0 & 0 & 0 & 0 & 0 \\ 0\,0\,0\,0\,0\,0\,0\,1\,0\,0\,0\,1\,1\,0\,0 & 1 & 1\,0\,1 & 0 & 0 & 0 & 0 & 0 \\ 1\,0\,0\,0\,0\,0\,0\,0\,0\,0\,0\,0\,0\,1\,0 & 0 & 0\,0\,0 & 1 & 0 & 0 & 0 & 0 \\ 0\,1\,0\,0\,0\,0\,0\,0\,0\,0\,0\,0\,0\,0\,0 & 0 & 0\,1\,0 & -1 & -1 & 0 & 0 & 0 \\ 0\,0\,1\,0\,0\,0\,0\,0\,0\,0\,0\,0\,0\,0\,1 & 0 & 0\,0\,0 & 0 & 1 & -1 & 0 & 0 \\ 0\,0\,0\,1\,0\,0\,0\,0\,0\,0\,0\,0\,0\,0\,0 & -1 & 0\,0\,0 & 0 & 0 & 1 & -1 & 0 \\ 0\,0\,0\,0\,1\,0\,0\,0\,0\,0\,0\,0\,0\,0\,0 & 0 & 0\,0\,-1 & 0 & 0 & 0 & 1 & 1 \\ 0\,0\,0\,0\,0\,1\,0\,0\,0\,0\,0\,0\,0\,0\,0 & 0 & -1\,0\,0 & 0 & 0 & 0 & 0 & -1 \end{pmatrix}$$

The implementation of the numerical procedure whose flowchart is given in Fig. 3.3 enabled the prediction of the fluxes of the different branches. Figure 3.7 shows the MEC results yielding the air gap flux density in the mover positions corresponding to the seven air gap reluctances. For the sake of validation, the waveform of the air gap flux density has been computed by a 2D FEA. The obtained results are illustrated in Fig. 3.7. They clearly highlight the agreement between the MEC and FEA models.

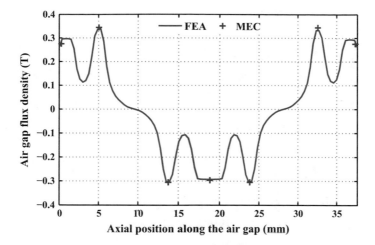

Fig. 3.7 Flux density, in given air gap positions, predicted by the MEC and validated by FEA

Table 3.1 Coil flux linkages yielded by the MEC and validated by FEA

Coils	MEC results (mWb)	FEA results (mWb)
ϕ_1	−0.0467	−0.0461
ϕ_2	−0.112	−0.112
ϕ_3	−0.0542	−0.0545
ϕ_4	0.0542	0.0551
ϕ_5	0.112	0.113
ϕ_6	0.0467	0.0479

Table 3.2 Phase flux linkages yielded by the MEC and validated by FEA

Phase	MEC results (mWb)	FEA results (mWb)
a	−0.101	−0.101
b	−0.101	−0.102
c	0.224	0.225

The fluxes linking the six coils are those crossing the yoke reluctances, noted $\phi_1, \phi_2, ..., \phi_6$ in Fig. 3.6. Table 3.1 gives the values of the flux linkages of the six coils, predicted by the MEC and computed by FEA.

Considering the winding distribution shown in Fig. 3.5, the flux linking each phase is predicted by the MEC as a sum of the flux linkages of the corresponding coils. The resulting fluxes as well as those computed by FEA are provided in Table 3.2. One can notice a good agreement between the MEC and FEA results.

Basically, in three phase rotating machines, the sum of the phase flux linkages at any rotor position is equal to zero, yielding the so-called "symmetrical winding". In linear machines, the phase flux linkages are basically-unbalanced. This statement is confirmed in the case of the studied machine. Indeed and referring to Table 3.2, the sum of the phase flux linkages is almost 0.022mWb which represents a relative error higher than 10%.

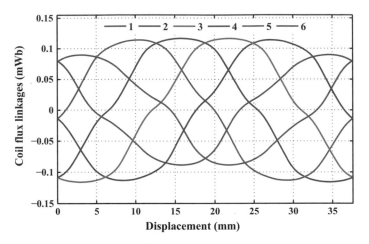

Fig. 3.8 Coil flux linkages of the studied IPM T-LSM

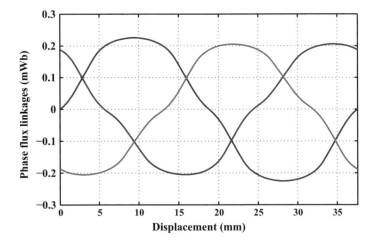

Fig. 3.9 Phase flux linkages of the studied IPM T-LSM

A FEA-based investigation of this shortcoming has been carried out. It consisted in the computation of the flux linkages of the six coils and then the ones of the three phases. The obtained results are shown in Figs. 3.8 and 3.9, respectively.

The analysis of Figs. 3.8 and 3.9 clearly highlights the asymmetrical coil flux linkages as well as the phase ones. Such an asymmetry concerns both angular phases and amplitudes. Obviously, this would has negative impacts on the machine outputs. These limitations are caused by the end effect phenomenon which is due to the linear machines open ends. Indeed, unlike rotating machines, linear ones suffer from the problem of the end effect which is more or less significant depending on the machine concept.

3.4 End Effect Cancellation

3.4.1 Study Statement

The end effect cancellation has been and continue to be a state of the art topic which has been widely-treated in the literature.

In [22], *Alonge* et al. proposed an approach to predict the parameters of the equivalent circuit of linear induction machines. In order to account for the end effect, they established a factor depending on the length of the mover, the electric time constant, the mover linear speed, and the air-gap thickness. Moreover, they developed an experimental procedure which enabled the prediction of the linear induction machine magnetic parameters and their variations with respect to the end effect.

In [23], *Accetta* et al. proposed a technique for the compensation of the static end effect in linear induction motor. This approach consisted in the addition of a synchronous controller in conjunction with a neural adaptive filter in order to extract the inverse sequence current component from the inductor current signature. It has been shown that, using the proposed synchronous controller, it is possible to control the negative current sequence component to zero.

In [24], *Lu* et al. analyzed the dynamic longitudinal end effect in long-primary double sided linear induction motors. An analytical model and an equivalent circuit of the studied machine has been established considering both the end effect by the use of correction factors. They found out that the influence of the end effect could be neglected at high speed operation and could reduced by increasing of the pole number of the secondary. The established results have been validated by 3D-FEA and by experiments.

In [16], *Xu* et al. proposed two-axis equivalent circuits of single-sided linear induction motors. They have found that the most influenced parameters by the edge end effect are the mutual inductance and the mover resistance. This statement has been accounted for, in the proposed models, by considering relative coefficients. The proposed two-axis equivalent circuit has been validated by experiments considering both steady-state operation and transient behavior of a prototyped linear induction motor.

In [25], *Cupertino* et al. proposed a position estimation scheme dedicated to linear low-saliency PM motors. The proposed scheme is based on the injection of a high-frequency current along the d-axis and on the analysis of the q-axis current. It has been shown that the proposed position observer exhibits a high accuracy, allied to an excellent robustness against the inverter nonlinearities, and to a cancelation of the end effect.

The same authors proposed, in a later work [26], an approach to compensate the end effect penalizing the performance of a tubular linear PM machine, under a sensorless position control strategy. In order to improve the accuracy of the position estimator, the end effect has been taken into consideration in the motor model.

In [27], *Hu* et al. proposed an analytical model of linear permanent magnet synchronous machines enabling the prediction of the magnetic field sliding in the air gap and the developed forces taking into account the primary end effect and the secondary one. The sub-domain approach has been integrated in the implemented analytical model, in order to compute the magnetic field, the tangential thrust and the normal forces. The obtained results have been validated by FEA and by experiments.

In [28], *Gruber* et al. proposed an approach to reduce the detent forces of a high force tubular PM linear synchronous machine (PMLSM). The approach consists in (i) adopting closed slots and in (ii) adding auxiliary poles at the end of the machine. In order to find out an optimal size and locations of the additional poles, a genetic algorithm combined with a FEA procedure have been considered. They have found that the influence of the additional poles on the end effect is independent from the machine length and have proved a reduction of the detent force of more than 50%.

In [29], *Qian Wang* and *Jiabin Wang* proposed an analysis and an assessment of motor-design-based techniques dedicated to the minimisation of the cogging forces in fractional-slot linear PM motors with either single-layer or double-layer non-overlapping windings. They have found that significant difference in cogging force characteristics exists between the motors with single-layer and double-layer non-overlapping windings.

Indeed, it has been shown that the cogging force of the linear PM motors with single-layer non-overlapping windings could be effectively reduced to a very low level, adopting appropriate design technics. Consequently, they are generally better suited than their double-layer counterparts for applications where smooth thrust force is critical. In contrast, linear PM motors with double-layer non-overlapping windings exhibit considerably higher tooth-ripple cogging forces even with the adopted fractional-slot design technique which, in turn, causes more difficulties in reducing the total cogging force at design stages. The effectiveness of the proposed cogging force reduction techniques have been confirmed through experiments carried out on a 14 pole 12 slot 3 phase linear PM motor prototype.

In [30, 31], *Souissi* et al. have proposed two approaches for the minimization of the end effect of the studied IPM T-LSM which are developed hereunder.

3.4.2   Proposed Approaches

The design of the studied IPM T-LSM has been rethought in an attempt to minimize the impact of the end effect on its electromagnetic features. Two approaches have been investigated. The corresponding introduced design changes are described and referred to the initial concept in Fig. 3.10.

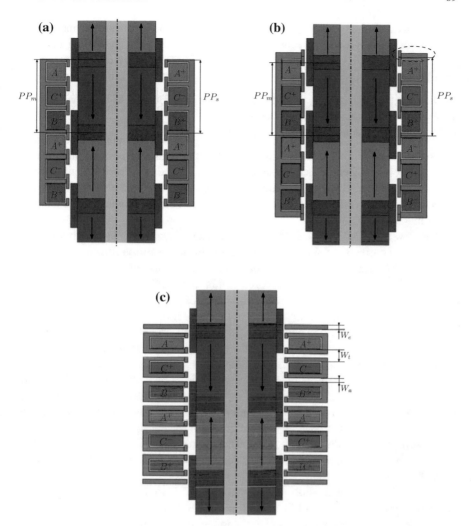

Fig. 3.10 Principle of the design approaches dedicated to the minimization of end effect of the studied IPM T-LSM. **Legend: a** initial concept, **b** concept rethought according to the first approach, **c** concept rethought according to the second approach

3.4.2.1   First Approach

The IPM T-LSM design has been rethought considering the following changes [30]:

- arranging the ratio R_P of the stator length L_s to the mover pole pitch τ, in order to achieve $\frac{2\pi}{3}$-shift between the phase flux linkages,
- extending the stator magnetic circuit with teeth of appropriate dimensions, characterized by the ratio R_T of the end tooth opening L_{tse} to the tooth opening L_{ts}, in order to balance the amplitudes of the phase flux linkages.

Following these changes, an update of the MEC model of the initial concept has been carried out and has led to the reluctance network shown in Fig. 3.11.

Referring to Fig. 3.11, one can notice that, unlike the MEC of the initial concept, the updated one incorporates:

- four poles instead of two, in order to cover the stator length when varying R_P,
- two more reluctances of the end teeth,
- an increase of the leakage fluxes to account for the ones located at the stator ends.

For the sake of determination of the best combination (R_P, R_T), enabling the minimization of the end effect, the established MEC has been solved while varying these ratios and keeping constant the mover geometry. The variation of (R_P, R_T) is associated to the changes of several reluctances especially those characterizing the stator magnetic circuit.

The developed MEC-based numerical procedure has been focused on the investigation of the effects of R_P and R_T on the sum of the phase flux linkages to the maximum flux ratio, the so-called "flux symmetry ratio" noted R_F. It should be underlined, that ideally, a null R_F corresponds to symmetrical phase flux linkages gained thanks to a total cancellation of the end effect.

The flowchart of the developed numerical procedure, enabling the resolution of the MEC of the IPM T-LSM under variable ratios R_P and R_T, is shown in Fig. 3.12.

Figure 3.13 illustrates the variation of the flux symmetry ratio R_F with respect to R_P and R_T. It clearly highlights the existence of a set of (R_P, R_T) combinations for which the ratio R_F is minimum. Furthermore, it is to be noted that for high values of R_P, and in spite of the increase of the phase flux linkages gained thanks to the reduction of the reluctances of the stator magnetic circuit, this latter turns to be bulky. Such a drawback turns to be relatively insignificant for R_P sightly higher than 100%.

To sum up, among the (R_P, R_T) combinations for which the IPM T-LSM exhibits a low value of R_F with acceptable size of the stator magnetic circuit ($100\% < R_P < 110\%$), one can select ($R_P = 108.5\%$, $R_T = 35\%$).

A FEA-based investigation of the machine no-load features, considering the selected combination ($R_p = 108.5\%$, $R_T = 35\%$), enabled the prediction of the coil and phase flux linkages, with respect to the mover displacement. The resulting waveforms are illustrated in Figs. 3.14 and 3.15, respectively. The analysis of the latter clearly highlights that the phase flux linkages turn to be almost symmetrical in terms of angular shifting and magnitude, indicating a quasi-cancellation of the end effect. However, it reveals that their harmonic content is not limited to the fundamental component, as stated in [30].

3.4.2.2 Second Approach

As shown in Fig. 3.10c, the second approach consists in assembling several modular units to build up the stator magnetic circuit, yielding the so-called "modular IPM T-LSM" [31]. Each unit is made up of a circumferential C-shaped core to which is associated a ring coil wound in the slot area. In order to decouple the units, flux

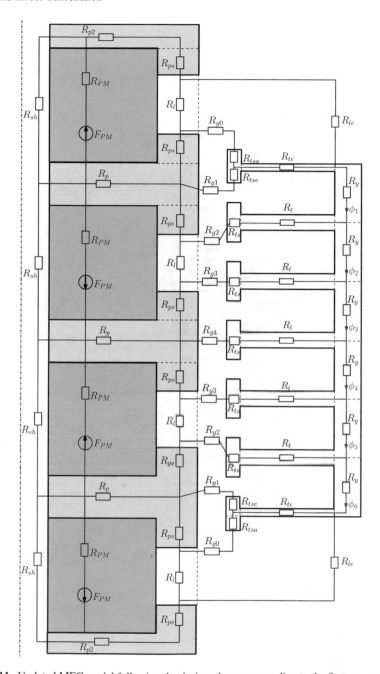

Fig. 3.11 Updated MEC model following the design changes according to the first approach

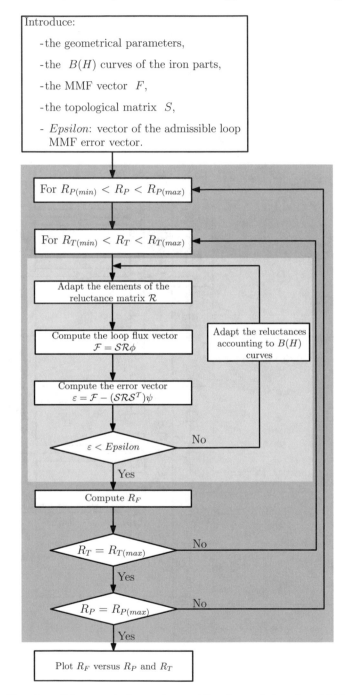

Fig. 3.12 Flowchart of the numerical procedure enabling the resolution of the MEC of the IPM T-LSM under variable ratios R_P and R_T

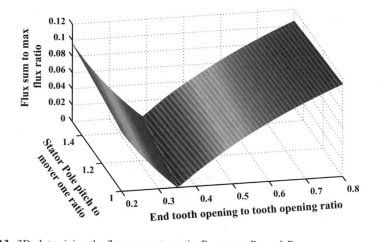

Fig. 3.13 3D plots giving the flux symmetry ratio R_F versus R_P and R_T

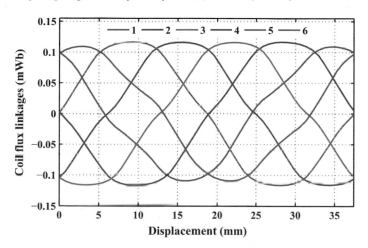

Fig. 3.14 Coil flux linkages of the IPM T-LSM corresponding to ($R_P = 108.5\%$, $R_T = 35\%$)

barriers, made up of a nonmagnetic material, are inserted in between adjacent C-shaped cores. Two core-discs are mounted in both extremities of the stator magnetic circuit.

Accounting for the main and major leakage fluxes, a MEC of the modular IPM T-LSM, inspired from the initial concept one, has been established considering the case where the mover is located at the position of maximum flux linkage with c-phase. The resulting MEC is illustrated in Fig. 3.16 where the blue and the red branches correspond to the end-tooth and to the flux barriers, respectively.

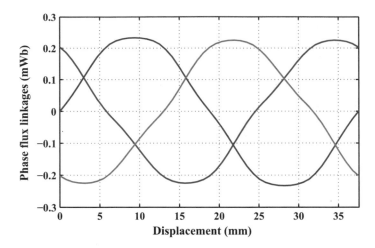

Fig. 3.15 Phase flux linkages of the IPM T-LSM corresponding to ($R_P = 108.5\%$, $R_T = 35\%$)

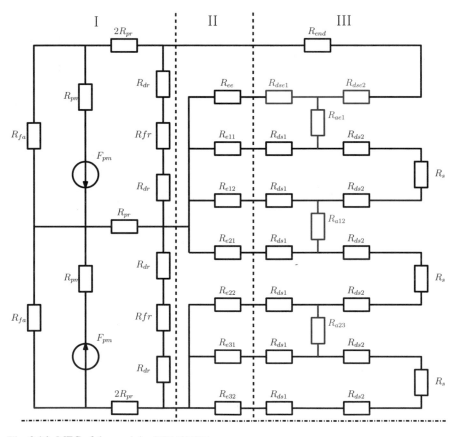

Fig. 3.16 MEC of the modular IPM T-LSM

Referring to the legend of Fig. 3.10c, two geometrical parameters have been selected in order to characterize the modular T-LSM, namely: W_a and W_e. These have been referred to parameters related to the stator tooth geometry through the definition of the following ratios:

$$\begin{cases} R_A = \dfrac{W_a}{W_t} \\[2mm] R_E = \dfrac{W_e}{W_{st}} \end{cases} \tag{3.9}$$

where W_t is identified in Fig. 3.10c and where W_{st} is defined as follows:

$$W_{st} = \frac{W_t - W_a}{2} \tag{3.10}$$

The sizing of the parameters W_a and W_e through the investigation of the effects of ratios R_A and R_E on the modular T-LSM, has been integrated in the resolution procedure of the MEC shown in figure staticmec. Such a procedure is similar to the one developed in the first approach with the substitution, in the flowchart given in Fig. 3.12, ratios R_P and R_T by ratios R_A and R_E. Two features allied to the end effect have been predicted while varying ratios R_A and R_E, which are:

- the sum of the phase flux linkages to the maximum flux (in c-phase) ratio, also named flux symmetry ratio, noted R_F,
- the stator tooth flux density, noted B_T.

The obtained results has led to the 3D plots illustrated in Fig. 3.17. As expected, for high values of R_A, the stator teeth turn to be saturated with a flux density reaching 2 T at no-load operation. Obviously, such a saturation is not in favor of the machine electromagnetic features. However, the end effect is no longer penalizing the phase flux linkages which turn to be almost balanced. Taking into consideration the previous

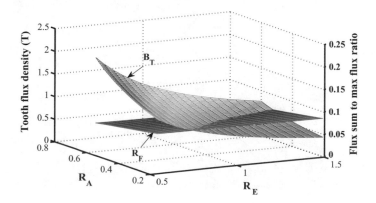

Fig. 3.17 3D plots giving the flux symmetry ratio R_F and the tooth flux density B_T versus ratios R_A and R_E

remarks, it clearly appears that there is a tradeoff in selecting R_A and R_E. Indeed, the two 3D plots exhibits a set of shared combinations (R_A, R_E) characterized by low values of R_f with an acceptable level of saturation, such as: $(R_A = 33\%, R_E = 100\%)$.

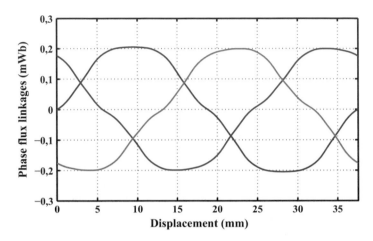

Fig. 3.18 Phase flux linkages of the modular T-LSM corresponding to $(R_A = 33\%, R_E = 100\%)$

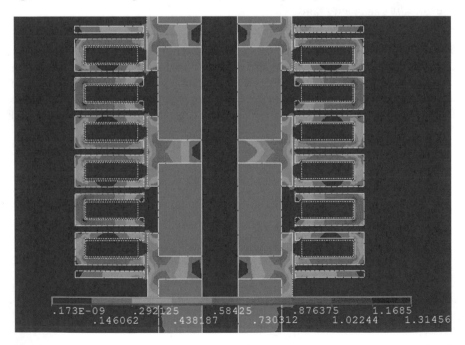

Fig. 3.19 Flux density mapping of the modular T-LSM corresponding to $(R_A = 33\%, R_E = 100\%)$

A FEA-based investigation of the no-load features of the modular T-LSM concept, with ($R_A = 33\%$, $R_E = 100\%$), has been carried out. It has led to the phase flux linkages shown in Fig. 3.18 which are quite balanced indicating the effectiveness of the modular design for the minimization of the end effect. The FEA has been extended to the investigation of the level of saturation. This has been achieved by the computation of the flux density mapping illustrated in Fig. 3.19. It is to be noted that the flux density does not exceed 1.5 T in the stator C-shaped cores, for the combination ($R_A = 33\%$, $R_E = 100\%$).

3.5 Incorporating the Mover Position

3.5.1 Background: Coupling Function

Basically, it is commonly admitted that MEC models enable the prediction of the machine features with an interesting compromise (accuracy/CPU time). This makes them suitable for machine pree-design procedures. However, MEC models are established for a given mover (rotor) position which limits their results. In order to extend their validity to the prediction of the time-varying features, the mover (rotor) position has to be incorporated in the MEC model, yielding the so-called "dynamic MEC".

Incorporating the rotor position in the MEC models of rotating machines has been treated by several teams [32–35] which is not the case of linear machines. The approach developed in [30, 36] is based on the following considerations:

- the reluctances characterizing the mover and stator (regions I and III) are kept constant when varying the mover position,
- the reluctances characterizing to the air gap (region II) are varied with the mover position according to the principle illustrated in Fig. 3.20.

The position varying reluctances are expressed as follows [30]:

$$R(\alpha) = \frac{1}{\mu_0} \frac{1}{2\pi z} \ln \left(\frac{R_g}{R_m} \right) \frac{1}{\alpha(z)} \tag{3.11}$$

where:

$$\begin{cases} \diamond \ R_m \text{ and } R_g \text{ are the air gap inner and middle radii, respectively,} \\ \diamond \ z \text{ is an axial length depending on the involved stator node,} \\ \diamond \ \alpha(z) \text{ is a mover-stator coupling function which varies with the mover displacement } z. \end{cases}$$

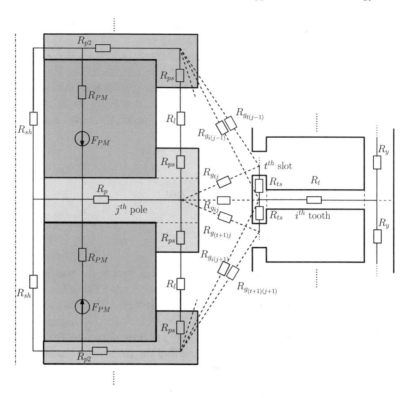

Fig. 3.20 Principle of the mover position incorporation in the MEC of the IPM T-LSM

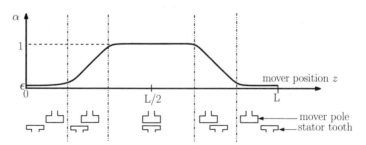

Fig. 3.21 Waveform of the sigmoid coupling function $\alpha(z)$

Referring to [30], it has been clearly shown that an interesting accuracy of the dynamic MEC could be gained thanks to the implementation of a sigmoid coupling function whose waveform has been establish according to the mover poles-stator teeth relative positions, as illustrated in Fig. 3.21.

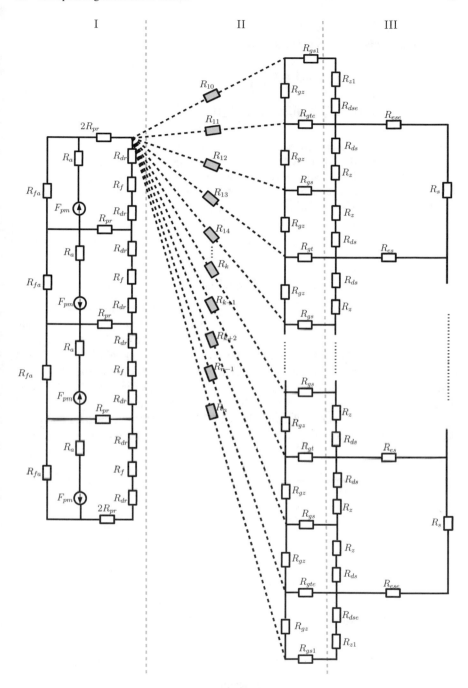

Fig. 3.22 Proposed dynamic MEC of the IPM T-LSM

An approach to formulate the sigmoid coupling function $\alpha(z)$, taking into account the machine geometry, has led to the following expression:

$$\alpha(z) = \begin{cases} \dfrac{1-\epsilon}{1-e^{(a(b-z))}} + \epsilon & \text{for} \quad 0 \le z \le \dfrac{L}{2} \\ \dfrac{1-\epsilon}{1-e^{(a(b-(L-z)))}} + \epsilon & \text{for} \quad \dfrac{L}{2} \le z \le L \end{cases} \tag{3.12}$$

$\begin{cases} \diamond & L \text{ is the sum of the stator slot and the mover pole openings,} \\ \diamond & a \text{ and } b \text{ are positive constants depending on the machine geometry which are estimated} \\ & \text{considering the slope and the delay, respectively, of } \alpha(z) \text{ shown in Fig. 3.21,} \\ \diamond & \epsilon \text{ is a small positive constant } (0 < \epsilon \ll 1) \text{ which has been introduced in order to avoid} \\ & \text{the division by "0" in Eq. (3.11).} \end{cases}$

3.5.2 Application to the IPM T-LSM

Incorporating the mover position, along with the end effect minimization according to the first design approach treated in Sect. 3.4.2.1, has led to MEC shown in Fig. 3.22.

The resolution of the dynamic MEC has been achieved by extending the numerical iterative procedure whose flowchart is given in Fig. 3.12, with the incorporation of an external loop dedicated to the mover position variation. With this done, the prediction of the time-varying features of the IPM T-LSM turns to be feasible.

Figures 3.23 and 3.24 give the waveforms of the coil and the phase flux linkages predicted by the proposed dynamic MEC, respectively. The corresponding FEA

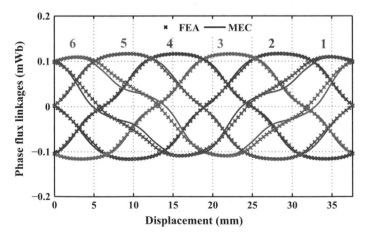

Fig. 3.23 Coil flux linkages of the IPM T-LSM predicted by the proposed dynamic MEC and validated by FEA

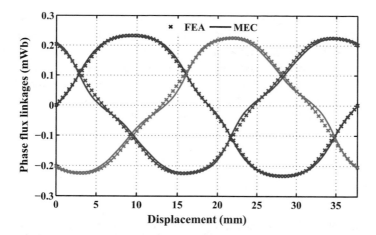

Fig. 3.24 Phase flux linkages of the IPM T-LSM predicted by the proposed dynamic MEC and validated by FEA

results have been plotted by dots in the same figures. One can notice a quite good agreement between the MEC and FEA results.

3.6 Conclusion

The third chapter highlighted the suitability of tubular linear synchronous machines (T-LSMs) to equip wave energy converters (WECs) which are currently considered as a state of the art topic. The integration of T-LSMs in such emergent sustainable application is mainly motivated by their improved energy efficiency compared to the one of WECs built around rotating generators.

The chapter has been initiated by a classification of WECs according to the technology of their power take-off systems with emphasis on the topology of the integrated generator: (i) rotating and (ii) linear topologies. Concerning rotating generators, doubly-fed induction and permanent magnet synchronous machines have been found viable candidates. As regards linear generators, variable reluctance and permanent magnet synchronous machines have been considered in several worldwide R&D projects dealing with wave energy harvesting with an interest by far greater in favor of the latter. Indeed, linear PM synchronous machines have been commonly integrated in direct drive WECs, thanks to their high force density and energy efficiency at low speeds. Of particular interest is the inset PM (IPM) tubular topology thanks to which an increase of the energy efficiency is gained with the elimination of the winding end turns so that the armature copper turns to be totally active. Furthermore, the radial attractive forces are diametrically-cancelled which limits the risk of eccentricity and reduces noise and vibration.

The second part of the chapter dealt with the MEC modelling of IPM tubular-linear synchronous machines (T-LSMs). Following, a recall of the general procedure dedicated to MEC resolution, it has been applied to the case of the IPM T-LSM. The preliminary results revealed the fallout of the end effect on the machine features. This made it necessary the rethought of the IPM T-LSM design according to two approaches whose effectiveness has been checked by FEA. Finally, the IPM T-LSM MEC validity has been extended to the time-varying features by incorporating the mover position in the machine model.

References

1. A. Clement, P. McCullen, A. Falcao, A. Fiorentino, F. Gardner, K. Hammarlund, G. Lemonis, T. Lewis, K. Nielsen, S. Petroncini, M.-T. Pontes, B.-O. Schild, P. Sjostrom, H.C. Soresen, T. Thorpe, Wave energy in Europe: current status and perspectives. Renew. Sustain. Energy Rev. **6**(5), 405–431 (2002)
2. B. Drew, A.R. Plummer, M.N. Sahinkaya, A review of wave energy converter technology. J. Power Energy **223**(part A), 887–902 (2009)
3. T. Ahmed, K. Nishida, M. Nakaoka, Grid power integration technologies for offshore ocean wave energy, in *Proceeding of the IEEE Energy Conversion Congress and Exposition*, Phoenix, Arizona, USA, Sept 2010, pp. 2378–2385
4. N. Muller, S. Kouro, J. Glaria, M. Malinowski, Medium-voltage power converter interface for wave dragon wave energy conversion system, in *Proceedings of the IEEE Energy Conversion Congress and Exposition*, Pittsburgh, Pennsylvania, USA, Sept 2013, pp. 352–358
5. N. Ahmed, M. Mueller, Impact of varying clearances for the wells turbine on heat transfer from electrical generators in oscillating water columns, in *Proceedings of the 2013 Eighth International Conference and Exhibition on Ecological Vehicles and Renewable Energies (EVER)*, Monte-carlo, Monaco, March 2013, pp. 1–6
6. H. Polinder, M. Mueller, M. Scuotto, M.G. de Sousa Prado, Linear generator systems for wave energy conversion, in *Proceedings of the European Wave and Tidal Energy Conference*, Porto, Portugal, Sept 2007, pp. 1–8
7. M. Blanco, M. Lafoz, G. Navarro, Wave energy converter dimensioning constrained by location, power take-off and control strategy, in *Proceedings of the IEEE International Symposium on Industrial Electronics (ISIE)*, Hangzhou, China, May 2012, pp. 1462–1467
8. J. Du, D. Liang, L. Xu, Q. Li, Modeling of a linear switched reluctance machine and drive for wave energy conversion using matrix and tensor approach. IEEE Trans. Magn. **46**(6), 1334–1337 (2010)
9. J. Pan, Y. Zou, G. Cao, Investigation of a low-power, double-sided switched reluctance generator for wave energy conversion. IET Renew. Power Gener. **7**(2), 98–109 (2013)
10. D. Wang, X. Wang, C. Zhang, Performance analysis of a high power density tubular linear switch reluctance generator for direct drive marine wave energy conversion, in *Proceedings of the International Conference on Electrical Machines and Systems (ICEMS)*, Hangzhou, China, Oct 2014, pp. 1781–1785
11. J. Du, P. Lu, X. Yang, Analysis and modeling of mutually coupled linear switched reluctance machine with transverse flux for wave energy conversion, in *Proceedings of the 2016 Eleventh International Conference on Ecological Vehicles and Renewable Energies (EVER)*, Monte-Carlo, Monaco, Apr 2016, pp. 1–6
12. D. Wang, C. Shao, X. Wang, C. Zhang, Performance characteristics and preliminary analysis of low cost tubular linear switch reluctance generator for direct drive WEC. IEEE Trans. Appl. Supercond. **26**(7), 0612205(1–5) (2016)

13. J.S. Kim, J.Y. Kim, S.K. Song, J.B. Park, The development of detent force minimizing permanent magnet linear generator for direct-drive wave energy converter, in *Proceedings of the 2012 Oceans*, Hampton Roads, Virginia, USA, Oct 2012, pp. 1–7

14. H. Polinder, M.E.C. Damen, F. Gardner, Linear pm generator system for wave energy conversion in the AWS. IEEE Trans. Energy Convers. **19**(3), 583–589 (2004)

15. J. Zhang, H. Yu, Q. Chen, M. Hu, L. Huang, Q. Liu, Design and experimental analysis of AC linear generator with halbach PM arrays for direct-drive wave energy conversion. IEEE Trans. Appl. Supercond. **24**(3), 0502704(1–4) (2014)

16. W. Xu, J.G. Zhu, Y. Zhang, Z. Li, Y. Li, Y. Wang, Y. Guo, Y. Li, Equivalent circuits for single-sided linear induction motors. IEEE Trans. Ind. Appl. **46**(6), 2410–2423 (2010)

17. O. Farrok, M.R. Islam, M.R.I. Sheikh, Y. Guo, J. Zhu, W. Xu, A novel superconducting magnet excited linear generator for wave energy conversion system. IEEE Trans. Appl. Supercond. **26**(7), 5207105(1–5) (2016)

18. C.A. Oprea, C.S. Martis, K.A. Biro, F.N. Jurca, Design and testing of a four-sided permanent magnet linear generator prototype, in **Proceedings of the International Conference on Electrical Machines (ICEM)**, Rome, Italy, Sept 2010, pp. 1–6

19. V. DelliColli, P. Cancelliere, F. Marignetti, R. DiStefano, M. Scarano, A tubular-generator drive for wave energy conversion. IEEE Trans. Industr. Electron. **53**(4), 1152–1159 (2006)

20. J. Prudell, M. Stoddard, E. Amon, T.K.A. Brekken, A. von Jouanne, A permanent-magnet tubular linear generator for ocean wave energy conversion. IEEE Trans. Ind. Appl. **46**(6), 2392–2400 (2010)

21. L. Cappelli, F. Marignetti, G. Mattiazzo, E. Giorcelli, G. Bracco, S. Carbone, C. Attaianese, Linear tubular permanent-magnet generators for the inertial sea wave energy converter. IEEE Trans. Ind. Appl. **50**(3), 1817–1828 (2014)

22. F. Alonge, M. Cirrincione, F.D. Ippolito, M. Pucci, A. Sferlazza, Parameter identification of linear induction motor model in extended range of operation by means of input-output data. IEEE Trans. Ind. Appl. **50**(2), 959–972 (2014)

23. A. Accetta, M. Pucci, A. Lidozzi, Compensation of static end effects in linear induction motor drives by frequency-adaptive synchronous controllers, in *Proceedings of the International Conference on Electrical Machines (ICEM)*, Berlin, Germany, Sept 2014, pp. 716–723

24. J. Lu, W. Ma, Research on end effect of linear induction machine for highspeed industrial transportation. IEEE Trans. Plasma Sci. **39**(1), 116–120 (2011)

25. F. Cupertino, G. Pellegrino, P. Giangrande, L. Salvatore, Sensorless position control of permanent-magnet motors with pulsating current injection and compensation of motor end effects. IEEE Trans. Ind. Appl. **47**(3), 1371–1379 (2011)

26. F. Cupertino, P. Giangrande, G. Pellegrino, L. Salvatore, End effects in linear tubular motors and compensated position sensorless control based on pulsating voltage injection. IEEE Trans. Industr. Electron. **58**(2), 494–502 (2011)

27. H. Hu, J. Zhao, X. Liu, Y. Guo, Magnetic field and force calculation in linear permanent-magnet synchronous machines accounting for longitudinal end effect. IEEE Trans. Industr. Electron. **63**(12), 7632–7643 (2016)

28. S. Gruber, C. Junge, R. Wegener, S. Soter, Reduction of detent force caused by the end effect of a high thrust tubular PMLSM using a genetic algorithm and FEM, in *Proceedings of the IEEE Annual Conference on Industrial Electronics Society (IECON)*, Glendale, Arizona, USA, Nov 2010, pp. 968-973

29. Q. Wang, J. Wang, Assessment of cogging-force-reduction techniques applied to fractional-slot linear permanent magnet motors equipped with non-overlapping windings. IET Electr. Power Appl. **10**(8–9), 697–705 (2016)

30. A. Souissi, M.W. Zouaghi, I. Abdennadher, A. Masmoudi, MEC-based modeling and sizing of a tubular linear PM synchronous machine. IEEE Trans. Ind. Appl. **51**(3), 2181–2194 (2015)

31. A. Souissi, I. Abdennadher, A. Masmoudi, On the stator magnetic circuit design of tubular-linear PM synchronous machines: A comparison between three topologies, in *Proceedings of the International Conference on Sustainable Mobility Applications, Renewables and Technology (SMART)* (Kuwait-City, Kuwait, 2015), pp. 1–8

32. H. Bai et al., Incorporating the effects of magnetic saturation in a coupled-circuit model of a claw-pole alternator. IEEE Trans. Energy Convers. **22**(2), 290–298 (2007)
33. M.L. Bash, J.M. Williams, S.D. Pekarek, Incorporating motion in mesh-based magnetic equivalent circuits. IEEE Trans. Energy Convers. **25**(2), 329–338 (2010)
34. M.L. Bash, S. Pekarek, Analysis and validation of a populationbased design of a wound-rotor synchronous machine. IEEE Trans. Energy Convers. **27**(3), 603–614 (2012)
35. D. Elloumi, A. Ibala, R. Rebhi, A. Masmoudi, Lumped circuit accounting for the rotor motion dedicated to the investigation of the time-varying features of claw pole topologies. IEEE Trans. Magn. **51**(5), 8105108(1–8) (2015)
36. M.W. Zouaghi, I. Abdennadher, A. Masmoudi, Lumped circuit-based sizing of quasi-Halbach PM excited T-LSMs: application to free piston engines. IET Electr. Power Appl. **11**(4), 557–566 (2017)

Chapter 4
Flat-Linear Synchronous Machines: Application to MAGLEV Trains

Abstract The chapter deals with the investigation of the magnetically-levitated (MAGLEV) trains. To start with, a study statement is carried out with emphasis on an historical overview of such technology and on the classification of MAGLEV trains with respect to several criteria. Giving the fact that the main issue of MAGLEV technology is the levitation, a special attention is paid to the classification according to the suspension system. Doing so, the electromagnetic suspension (EMS) and electrodynamic suspension (EDS) systems are deeply investigated. Indeed, considering EMS two electromagnet configurations are studied which are: (i) flat track with U-shaped core electromagnet and (ii) U-shaped track with U-shaped core electromagnet, for which the produced magnetic forces and the state equations are derived. For the EDS technology, two systems are considered which are: (i) a moving coil over conducting sheet and (ii) moving coil facing figure-eight null-flux coil. Such concepts are studied with emphasis on the levitation as well as the guidance forces.

Keywords Magnetic levitation · MAGLEV train · Electromagnetic suspension
Electrodynamic suspension · Attractive magnetic forces
Repulsive magnetic forces

4.1 Introduction

Transportation is an important issue in each stage of human civilization. Indeed, it contribute to the development of economic, social, political and cultural fields. Giving this fact, without the development of the means of transport the world would be totally different. Doing so, remarkable improvements was, are and will be depicted in this field leading to a wide variety in the means of transport.

After the invention of wheel, trains are the subject of an endless progress from the steam powered locomotives reaching the high-speed train. However, wheels are no longer a necessity for the trains in the case of magnetically-levitated (MAGLEV) ones. In addition to the substitution of the propulsion system, MAGLEV technology eliminate wheels, gearboxes and bearings, resulting mainly on a very higher speed reaching up to 600 km/h.

© Springer Nature Singapore Pte Ltd. 2019
A. Souissi et al., *Linear Synchronous Machines*, Power Systems,
https://doi.org/10.1007/978-981-13-0423-1_4

MAGLEV trains differ from conventional ones in that they are levitated, guided and propelled along a specific guideway through an exchange of magnetic fields rather than by steam, diesel or electric engine. Indeed, the substitution of mechanical contacts by magnetic interaction offers the possibility to reach higher speeds (up to 600 km/h) and efficiencies.

In addition, the elimination of wheels and track wear enables a reduction of the maintenance frequency and costs in one hand, and of a minimization of noise, vibration, and risk of slipping leading to a higher safety and comfort in the other hand.

Moreover and compared to rivals, MAGLEV trains are considered as eco-friendly mean of transports as they are free of any emission of greenhouse gazes which are currently causing many climatic disasters due to the global worming phenomenon. This statement is stronger, if the consumed energy by these trains is generated mainly based on green renewable energies. Considering other aspects, MAGLEV trains present several superiorities versus conventional trains as well as airplanes, such that: (i) a minor affection by the weather (i.e. snow, ice, severe cold, rain or high winds), (ii) a competitive journey times over distances of 800 km or less.

Considering all these advantages, an increasing development is depicted in such field despite the fact that MAGLEV train improvement is braked by the high cost of the installation as well as the train. This drawback is mainly caused by the special track needed for the MAGLEV trains and their innovative aspect.

With this said, the present chapter is devoted to the analysis of the most used technologies of MAGLEV trains. To Start with a study statement focusing on the MAGLEV trains history, trends and classifications is carried out. Then, two basic suspension technologies that are: (i) Electromagnetic suspension (EMS) and (ii) Electrodynamic suspension (EDS), systems are investigated.

4.2 Study Statement

4.2.1 Historical Overview

Basically, MAGLEV is a way of suspension that levitate object without any contact by means of magnetic field. Such designation is commonly known as the name of the magnetically-levitated trains as it nowadays is the most famous application. However, their penetration to rail way transportation remains limited despite the fact that such technology was developed since the 1900's century.

Indeed, the American *Robert Goddard* and the French *Emile Bachelet* conceived the idea of frictionless trains. Its pathway is the seat of magnetic field generated by transverse and longitudinal series of electro-magnets arranged along the path. Such invention was designed for the transmission of mail or small express packages from a local to an other or for moving cash and parcel carriers in large department stores. The scientists research does not exceed the level of prototyping due to the lack of financial support. Doing so, MAGLEV concept evolution has been stopped for about 60 years until the Japanese and the German started to do research on the subject [1].

The first commercial MAGLEV train in the world was a low-speed MAGLEV shuttle installed in Birmingham, United Kingdom in 1984. Its track length was 600 m, and the cars was levitated at an altitude of 15 mm. The levitation was assured by electromagnets, and propulsion with linear induction motors [2].

The Japanese started their research on MAGLEV transportation in the beginning of the 1970's. After many years of experiments the Japanese constructed their first test line, 7 km in 1977. In 1990, Japan constructed the Yamanashi MAGLEV test line which has a 42.8 km long and has host the first running test in 1997. On 21 April 2015, the "L0" MAGLEV reaches a speed of 603 km/h which is the highest speed achieved up to now [3]. They also developed the world's first unmanned commercial urban MAGLEV called "Linimo" [4].

The Germanies started research on MAGLEV train in early 1980's in several countries. After several tests, their most known MAGLEV, "Transrapid" was installed in Shanghai, China in 2003 with a line of 30 km long. The train can reach 350 km/h in 2 min, with a maximum normal operation speed of 431 km/h [5].

Following several research, the Korean developed the world's second commercialized operating unmanned urban MAGLEV. It is called "ECOBEE" and is characterized by 6.1 km long, six stations and an operating speed of 110 km/h . Two more stages are planned of 9.7 km and 37.4 km . Once completed it will become a circular line [6].

4.2.2 MAGLEV Classification

MAGLEV trains could be classified considering several criteria as shown in Fig. 4.1:

- suspension system; It could based on attractive electromagnetic forces or on repulsive electrodynamic ones [7],
- propulsion motor type; Several linear motors could be integrated in MAGLEVs. These should have a flat geometry, but could be synchronous or induction motors [8],
- type of the guideway; According to the adopted technology and motor type, the guideway could host a part of an electric motor leading to the so-called active guideway. In the case of the passive guideway, all the active motor part are placed on the mover [9],
- range of speed; despite the fact that MAGLEV trains are known by their high speed MAGLEVs, they could be classified into low speed for operating speed lower than 100 km/h or high speed MAGLEVS,
- driving way; MAGLEV trains could be without driver leading to unmanned device.

Among these criteria the most commune one is the suspension system. Doing so, a deeper investigation of EMS and EDS systems will be carried out.

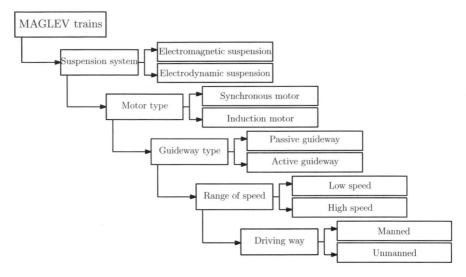

Fig. 4.1 Classification of MAGLEV trains

4.2.2.1 Electromagnetic Suspension

Electromagnetic suspension systems is based on attractive forces between current-controlled electromagnets and magnetic core. Basically, for MAGLEV trains the primary of the considered device is hosted on board the vehicle and is equipped by PM or controlled electromagnet. The secondary is fixed on the track and is made up of solid magnetic core. Doing so, the track is characterized by a T-shape and is wrapped by the undercarriage of the vehicle as illustrated in Fig. 4.1, where are placed the electromagnet. As they are on board the vehicle the electromagnet get easily their energy from battery pack installed on the train through an appropriate control system. As the levitation electromagnets are placed under the track, the attractive magnetic forces will attract the undercarriage of the vehicle leading to its levitation over the track [10] (Fig. 4.2).

EMS-based trains use mainly DC current-controlled electromagnets instead of PMs, as they should be continuously controlled depending on the load and the external disturbances. Doing so, several electromagnets configurations could be adopted. Moreover, the electromagnet could achieve the levitation as well as the guidance or only the levitation and thus is mainly according to the geometry of the track. Considering the last case an additional electromagnets should be installed to achieve the guidance of the train along the guideway.

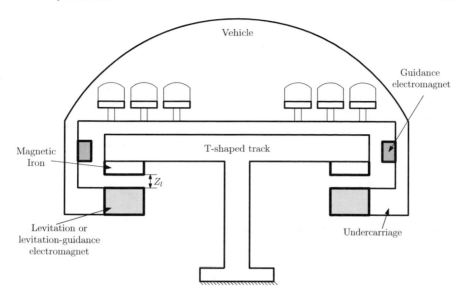

Fig. 4.2 EMS-based MAGLEV train concept

4.2.2.2 Electrodynamic Suspension

Electrodynamic suspension (EDS) systems is based on repulsive forces which are basically the result of the interaction between two magnetic fields of same polarity. Such interaction could be produced between, (i) two current carrying conductors of same polarity, (ii) two PMs of same polarity, (iii) a moving current carrying conductor and a fix electric circuit which could be a conducting sheet or a short-circuited coil, or (iv) a moving PM and a fixed short-circuited electric circuit.

Among the above cited method, the most integrated in the MAGLEV trains is the moving current carrying conductor and a fix electric circuit which could be a conducting sheet or a short-circuited coil. For such application, to obtain high levitation force per weight for an mechanical air gap around 100 mm, superconducting (SC) coils are required for their high delivered magnetic field [11]. However, to obtain the superconductivity phenomenon, the coils should be cooled under their critical temperature T_C (almost equal to $-150°$ C) which requires an appropriate cooling system. Doing so, the superconducting electromagnets should be placed on board the vehicle, facing a fixed conducting sheet or short-circuited coil mounted on the guideway as illustrated in Fig. 4.3. The technology shown in Fig. 4.3b is the one adopted by the Japanese in the "L0" MAGLEV train which have the record of the fastest train as it reaches a speed of 603 km/h in 2015. However, this train is till now in the testing level and it is not yet commercialized.

Considering PMs, an other technology which could be under the EDS type is the inductrack one. It integrates rare earth PMs mounted within a *Halbach* arrangement. Such array moves facing short-circuited conductors located in the track [12]. This

(a)   **(b)**

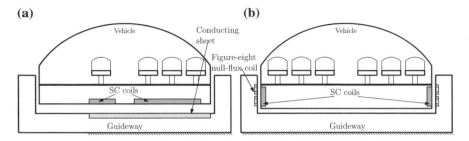

Fig. 4.3 EDS-based MAGLEV train of the most used concept. **Legend: a** moving SC coils over conducting sheet, **b** moving SC coils facing a short circuited coil

technology make it possible the elimination of the SC coils in order to achieve adequate levitating forces. And thus all the cooling system will be eliminated. However, such concept still in the prototyping step.

4.3 Electromagnetic Suspension Systems

4.3.1 Electromagnets Configurations

EMS-based MAGLEVs are based on DC current controlled electromagnets that could have several configurations. It depends on (i) the geometry of its core, (ii) the one of the back iron, or (iii) the arrangement of the coil [9]. Thus, the core of the electromagnet could be U- or E- shaped. However, the back iron could take flat or U-shape geometry. Several configurations are shown in Fig. 4.4.

EMS systems could be classified according to the geometry of the back iron (track), as this latter is very influencing in terms of achieved functions by the system. Indeed, if the back iron is flat, the corresponding EMS system is dedicated only for the levitation. However, if it adopt the U-shape with a U-shaped electromagnet core, such system ensure the levitation as well as the guidance.

As the EMS system equipped with flat track ensure exclusively the levitation, an additional electromagnets should be installed in the lateral sides of the MAGLEV train and thus in order to achieve the guidance of the vehicle as illustrated in Fig. 4.4. Such technology was adopted and developed by the Germanies leading to "Transrapid" MAGLEV train which is installed since 2004 in Shanghai. In addition to the levitation and guidance electromagnets, it is equipped by a long-stator linear synchronous motor for the propulsion with a rated speed of 420 km/h.

The U-shaped track electromagnets are mainly integrated in the "Linimo" and "ECOBEE" MAGLEV trains which was developed by the Japanese and the Korean, respectively. Both MAGLEVs are considering as a low-speed ones as they are designed to operate under a maximum speed of 100 km/h.

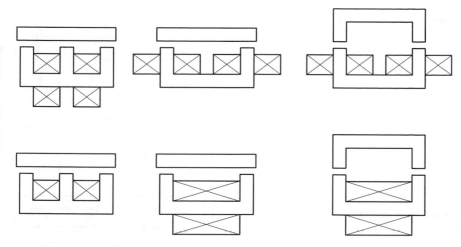

Fig. 4.4 Typical electromagnet geometries with different coil configuration

4.3.2   Flat Track-Based EMS System

Flat track-based EMS systems are considered as mono-function ones as they achieve only the levitation or the guidance and this depending on the emplacement of the electromagnet on the train. It could be installed in the undercarriage or in the lateral sides, respectively. Although the electromagnet achieves the levitation or the guidance, the study is almost similar, except the direction of the produced forces which could be selected according the electromagnet direction. Doing so, the MAGLEV train should be equipped at each stage by two independent electromagnets, one to ensure the levitation and an other to achieve the guidance.

Giving the fact that the U-core electromagnet is the most used in the MAGLEV trains, a special attention will be paid to the systems equipped by such electromagnet. Doing so, an electromagnet characterized by a flat track as shown in Fig. 4.5 will be studied with emphasis on the magnetic forces [13]. Based on the energy formulations and by considering an unidirectional force along the z-axis, the magnetic force is expressed based on Eqs. (2.82) and (2.100) as:

$$F = \frac{B^2}{2\mu}S \tag{4.1}$$

where B is the magnetic field reigning in a region characterized by a permeability μ crossing a surface S.

In our case the study will be concentrated on the air gap region leading to $\mu = \mu_0$ and S is the air gap section which could be expressed as $S = C_y L$, where L is the active length of the electromagnet along the x-axis.

Fig. 4.5 U-shaped core-flat track electromagnet

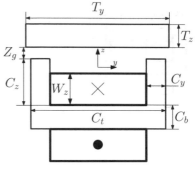

Fig. 4.6 MEC model of the studied electromagnet

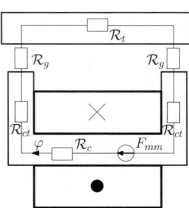

Giving the fact that such formulation is based on the prediction of the magnetic field, it will be highlighted hereunder.

4.3.2.1 Magnetic Field

Based on the MEC resolution procedure derived in Chap. 3, the flux density B could be determined through the prediction of the magnetic flux φ crossing the air gap surface S, then applying the following expression:

$$B = \frac{\varphi}{S} \qquad (4.2)$$

The determination of the flux φ will be done based on the resolution of a MEC. Doing so, the equivalent circuit associated to the studied electromagnet is shown in Fig. 4.6.

Referring to the geometrical parameters shown in Fig. 4.5 and considering Eq. (2.38), the reluctances $\mathcal{R}_t$, $\mathcal{R}_g$, $\mathcal{R}_{ct}$ and $\mathcal{R}_c$ could be expressed as:

$$
\begin{cases}
\mathcal{R}_t = \dfrac{C_t}{\mu_0 \mu_r T_z L} \\[2mm]
\mathcal{R}_g = \dfrac{Z_g}{\mu_0 C_y L} \\[2mm]
\mathcal{R}_{ct} = \dfrac{C_z}{\mu_0 \mu_r C_y L} \\[2mm]
\mathcal{R}_c = \dfrac{C_t}{\mu_0 \mu_r C_b L}
\end{cases}
\tag{4.3}
$$

Giving the fact that the relative permeability μ_r of the electromagnet core as well as of the track is higher compared to the one of the air (μ_r(air)= 1), their reluctances will be neglected in an attempt to simplify the study. Doing so and applying the *Kirchhoff*'s voltage law, one can deduce:

$$
2\varphi \mathcal{R}_g - F_{mm} = 0
\tag{4.4}
$$

where the magnetomotive force could be expressed as $F_{mm} = NI$.

Based on Eqs. (4.4) and (4.2), the magnetic field φ is formulated as:

$$
B_{MEC1} = \frac{\mu_0 N I}{2 Z_g}
\tag{4.5}
$$

In order to achieve a better accuracy, we could take into account at least the reluctance of the track R_t. Doing so, and considering that the width of the track is equal to the one of the teeth core ($T_z = C_y$), the magnetic field B_{MEC2} be could be expressed as:

$$
B_{MEC2} = \frac{\mu_0 \mu_r N I}{2 \mu_r Z_g + C_t}
\tag{4.6}
$$

Moreover and referring to Eqs. (4.5) and (4.6), one can deduce that taking into account all the MEC reluctances make it more complex the magnetic field expression and so the one of the force. In order to justify the fact that the iron parts reluctances are neglected, a case study will be considered and a comparison between the magnetic field Eqs. (4.5) and (4.6) will be carried out.

Let us consider that the studied electromagnet shown in Fig. 4.5 is characterized by the parameters given in Table 4.1.

The variation of the magnetic fields B_{MEC1} and B_{MEC2} with respect to the air gap Z_g, for a constant current $I = 20A$, is yielded by the resolution of Eqs. (4.5) and (4.6), respectively, and is illustrated in Fig. 4.7. Its analysis leads to:

- for low values of the air gap ($Z_g < 6\,mm$), the magnetic field is widely affected by the changes of Z_g,
- for bigger values of air gap ($Z_g > 8\,mm$), the magnetic fields present a slight variation within a limited range,
- the curves of the variation of B_{MEC1} and B_{MEC2} have a quasi-identic forms except an insignificant disparity for the low values of the air gap ($Z_g < 2\,mm$).

Table 4.1 Electromagnet parameters

Symbol	Value
C_t	0.15 m
C_b	0.03 m
W_z	0.025 m
C_y	0.04 m
L	1 m
μ_r	1000
N	100 turn

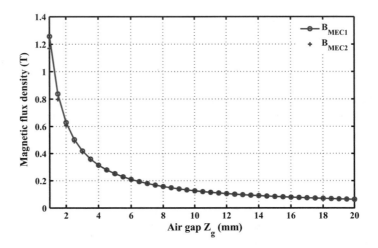

Fig. 4.7 Variation of the magnetic fields B_{MEC1} and B_{MEC2} with respect to the air gap Z_g

Accounting for the above-cited constatations and for the simplicity of Eq. (4.5), this latter will be considered in what follows.

4.3.2.2   Magnetic Force

The magnetic force F_m formulation will be predicted based on Eq. (4.1) and thus by the substitution of B given by Eq. (4.5) leading to:

$$F_m = \frac{\mu_0 N^2 S}{8} \frac{I^2}{Z_g^2} \tag{4.7}$$

Referring to Eq. (4.7), it clearly appears that for constant current values, the force increases with the decrease of the air gap and viceversa. So, more the electromagnet is closer to the track the bigger are the attractive magnetic forces. In order to highlight this statement, the magnetic force is computed using Eq. (4.7) considering the

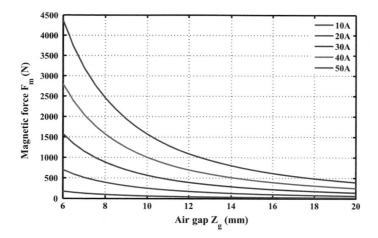

Fig. 4.8 Variation of the magnetic force F_m with respect to the air gap Z_g for different values of the current I

parameters given in Table 4.1 and for different current values. The resulting curves are shown in Fig. 4.8.

From the analyses of Fig. 4.8, one can deduce that the larger the air gap, the smaller is the distortion of the force dF_m and the lower is the produced force F_m. Thus, the selection of the adequate air gap should be based on a the compromise high magnetic force/low force distortion. Despite such selection, for constant current value the considered system is systematically unstable and can not maintain a constant air gap. Indeed, if an external disturbance force is applied to the system operating for an equilibrium air gap, the train will be adhered or separated from the track depending on the direction of the distortion. Such instability make it essential the incorporation of appropriate system to control the feeded current continuously. An adequate sensors should be adopted to detect all external disturbance and under the action of the control system, the attractive magnetic force is instantaneously adjusted.

4.3.2.3   Elementary Flat Track-Based EMS System

Let us consider an elementary EMS system based on flat track electromagnets. As these latter produce a unidirectional force, such system is equipped by two levitation electromagnets at the train undercarriage sides and two guidance electromagnet at the lateral train sides [9]. Accounting for the symmetry of the system only the half of the cross section is illustrated in Fig. 4.9, where the forces applied on the train are mentioned.

Fig. 4.9 Half cross section
of an EMS flat track
electromagnet based train

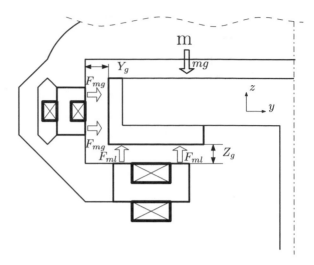

These forces are related by the Newton's second law; In an inertial reference
frame, the vector sum of the forces applied on an object is equal to the mass m of
that object multiplied by the acceleration $\vec{a}$ of the object; as:

$$\sum \vec{F} = m\,\vec{a} \tag{4.8}$$

Levitation Following the application of Eq. (4.8), the projection of the forces and
the acceleration on the z-axis leads to:

$$ma_z = F_M - mg \tag{4.9}$$

where:

- a_z could be written as the square derivative of the displacement z, $a_z = d^2z/dt^2 = \ddot{z}$,
- F_M is the magnetic force and is equal to $4F_{ml}$ which is expressed in Eq. (4.7),
- g is the gravity acceleration.

Doing so, Eq. (4.9) turn to be:

$$m\ddot{z} = C_l \frac{I^2}{Z_g^2} - mg \tag{4.10}$$

where the constant C_l depend of the electromagnet core geometry. Considering the
U-shaped core as shown in Fig. 4.9, such constant is expressed as:

$$C_l = \frac{\mu_0 N^2 S_l}{2} \tag{4.11}$$

It clearly appears that the differential equation derived from Eq. (4.9) is a non linear one. The resolution of such equation goes through its linearization with respect to both variables Z_g and I around the equilibrium point (I_e, Z_{ge}) which is the operating point at steady state and is characterised by:

$$\begin{cases} Z_g = Z_{ge} \\ I_e = \sqrt{\dfrac{mg}{C_I}} Z_{ge} \end{cases} \qquad (4.12)$$

The use of *Taylor* expansion for the linearization of the magnetic force F_m at the equilibrium point leads to:

$$F_M = C_z(Z_g - Z_{ge}) + C_I(I - I_e) \qquad (4.13)$$

where:

$$\begin{cases} C_z = \dfrac{\partial F_M}{\partial Z_g}|_{(Z_g=Z_{ge}, I=I_e)} \\ C_I = \dfrac{\partial F_M}{\partial I}|_{(Z_g=Z_{ge}, I=I_e)} \end{cases} \qquad (4.14)$$

It is to be underlined that such linearized form is valid only around the equilibrium point. So, if we consider an elementary distortion along the z-direction the linearized form of Eq. (4.10) will be as follows:

$$m\ddot{Z}_g + 2C\frac{I_e^2}{Z_{ge}^3}Z_g - 2C\frac{I_e}{Z_{ge}^2}I = 0 \qquad (4.15)$$

From the analysis of linearized state Eq. (4.15), it is to be underlined that the effect of the magnetic force, around the equilibrium point, on the displacement z as well as on the current is similar to the one of the spring of stiffness $k_Z = \dfrac{2C_I I_e^2}{Z_{ge}^3}$ and $k_I = \dfrac{2C_I I_e}{Z_{ge}^2}$, respectively. So, the magnetic force has the trend to oppose the variation in air gap Z_g and in current I. Doing so, such system is considered as an undamped mass-spring system. It is primordial to add damper to avoid the free vibration phenomenon. However, a deep frequency analysis should be carried out to avoid disturbance characterized by a resonant frequency. The appropriated speeds are considered as critical ones.

If we consider that the electromagnet coil is feeded by the current I through a resistor R, the current-voltage equation is as follows:

$$V = RI + \frac{d(LI)}{dt} = RI + \frac{dL}{dt}I + \frac{dI}{dt}L \qquad (4.16)$$

where L is the equivalent inductance of the electromagnet coil.

Referring to Eq. (2.84), the substitution of the loop equivalent reluctance deduced based on the air gap one R_g (system Eq. 4.14) leads to the expression of equivalent inductance L as:

$$L = \frac{C_l}{Z_g} \tag{4.17}$$

The substitution of Eq. (4.17) into Eq. (4.16) yields:

$$V = RI + \frac{C_l}{Z_g}\dot{i} - \frac{C_l I}{Z_g^2}\dot{Z}_g \tag{4.18}$$

In addition to the system (4.12), the equilibrium point is characterized by:

$$V_e = RI_e \tag{4.19}$$

Following the linearization of Eq. (4.18) with respect to I and Z_g around the equilibrium point, if we consider an elementary displacement around the equilibrium point the differential equation turns to be:

$$V = RI + \frac{C_l}{Z_{ge}}\dot{i} - \frac{C_l I_e}{Z_{ge}^2}\dot{Z}_g \tag{4.20}$$

Giving the fact that the electromagnet motion is within a small range near the equilibrium point (Z_{ge}, I_e), the electromagnet inductance may be approximated at a constant value as:

$$L = \frac{C_l}{Z_{ge}} \tag{4.21}$$

Thus, the Eq. (4.20) turns to be:

$$V = RI + \frac{C_l}{Z_{ge}}\dot{i} \tag{4.22}$$

If we combine the mechanical model and the electric one by considering Eqs. (4.15) and (4.22), we obtain the following system:

$$\begin{cases} \ddot{Z}_g = 2C_l \dfrac{I_e}{mZ_{ge}^2}I - 2C\dfrac{I_e^2}{mZ_{ge}^3}Z_g \\ \dot{i} = \dfrac{Z_{ge}}{C_l}V - \dfrac{Z_{ge}R}{C_l}I \end{cases} \tag{4.23}$$

Based on the above mentioned system (4.23), one can deduce the bloc diagram illustrated in Fig. 4.10.

Referring to Fig. 4.10, the relative air gap position, velocity and acceleration as well as the electromagnet flux ($\dot{I}$) are needed in the feedback loop to maintain the stability of the air gap and thus of the levitated train. Doing so, several configuration could be adopted, Among which the one illustrated in Fig. 4.11. Such configuration is mainly based on the association of the gap sensor, the controller, and the electromagnet [14].

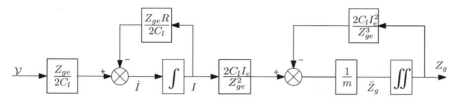

Fig. 4.10 Bloc diagram of the levitation system based on the linearized state equations

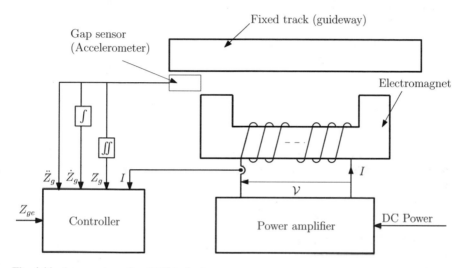

Fig. 4.11 An overview of an EMS levitation system

Guidance

Referring to Fig. 4.5, the lateral electromagnet achieving the guidance are in inter-action with the track through attractive magnetic forces F_{mg} with in the lateral air gap Y_{ge}. Without any external disturbance, the train guided along the guide way with a constant lateral air gap for both sides. Doing so and giving the fact that the two electromagnets are identic and feeded by the same current, the resultant magnetic force yielded by the set of the two lateral electromagnet is null as the left and right forces have the same value with apposite direction.

However, under an external disturbance modeled by the force f_d, a lateral dis-placement Δy will be caused as shown in Fig. 4.12. Thus, it leads to an unbalanced lateral air gap between both sides ($y_l \neq y_r$) and thus unequal magnetic forces. Doing so, the resultant lateral force opposed the disturbance one in order ensure the guidance of the vehicle along the guide way.

Following the projection of the applied forces on the y-axis, Eq. (4.8) leads to:

$$m\ddot{y} = f_d + F_{mgl} - F_{mgr} \tag{4.24}$$

where F_{mgl} and F_{mgr} are the forces of the left and right electromagnet, respectively.

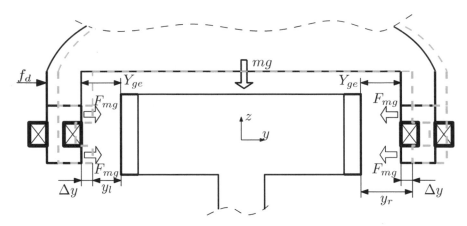

Fig. 4.12 Elementary guidance system

Giving the fact that both electromagnets have similar to the one studied in the previous paragraph, the left and right magnetic forces are expressed based on Eq. (4.7) as:

$$\begin{cases} F_{mgl} = \dfrac{\mu_0 N^2 S_g}{4} \dfrac{I^2}{y_l^2} = \dfrac{C_g}{2} \dfrac{I^2}{y_l^2} \\ F_{mgr} = \dfrac{\mu_0 N^2 S_g}{4} \dfrac{I^2}{y_r^2} = \dfrac{C_g}{2} \dfrac{I^2}{y_r^2} \end{cases} \tag{4.25}$$

where the left and right lateral air gaps y_l and y_r are given by, $y_l = Y_{ge} - \Delta y$ and $y_r = Y_{ge} + \Delta y$, as shown in Fig. 4.12, and where C_g is defined as C_l by Eq. (4.11).

The linearization of the above cited magnetic forces around an equilibrium point characterized by (Y_{ge}, I_e) and thus by $(\Delta y = 0)$, leads to the following equations:

$$\begin{cases} F_{mgl} = C_g \dfrac{I_e}{Y_{ge}^2} I - C_g \dfrac{I_e^2}{Y_{ge}^3} y_l \\ F_{mgr} = C_g \dfrac{I_e}{Y_{ge}^2} I - C_g \dfrac{I_e^2}{Y_{ge}^3} y_r \end{cases} \tag{4.26}$$

Taking into account the expressions of y_l and y_r in terms of Y_{ge}, a linearized form of Eq. (4.24) could be expressed as follows:

$$m\ddot{y} = f_d + 2C_g \frac{I_e^2}{Y_{ge}^3} \Delta y \tag{4.27}$$

Similarly to the levitation system, the current-voltage Eq. (4.22) is applied to the guidance electromagnets leading to the following linearized state equations:

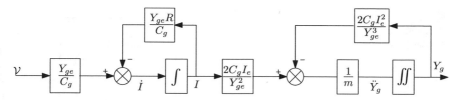

Fig. 4.13   Bloc diagram of the guidance system

$$\begin{cases} \ddot{y} = 2C_g \dfrac{I_e^2}{m Y_{ge}^3} \Delta y \\[3mm] \dot{I} = \dfrac{Y_{ge}}{C_g} \mathcal{V} - \dfrac{Y_{ge} R}{C_g} I \end{cases} \qquad (4.28)$$

Doing so, the corresponding bloc diagram is as illustrated as Fig. 4.13.

Similar to the levitation system, the guidance one is unstable system and should be continuously under control to keep constant the air gaps within the y-direction. The several disturbance forces within the y-axis could be predicted according to the profile of the guideway which make the electromagnet handling easier and predictable. It should be underlined that left and right electromagnets could be controlled separately.

The flat-shaped track electromagnet require the use of separated electromagnet for the guidance and the levitation. Thus, the number od controllers and sensors is twice high. To over come such limitation, an alternative solution is to adopt the U-shaped track electromagnet which ensure the levitation as well as the guidance. It is to be noted that the use of U-shaped electromagnet is integrated only in low speed MAGLEVs. This concept will considered hereunder.

4.3.3   U-Shaped Track-Based EMS System

U-shaped electromagnets with U-shaped track are adopted to achieve simultaneously the levitation and the guidance which greatly reduce the number of the electromagnet and thus all associated system such as control blocks and sensors [15]. However, such technology is mainly adopted for law-speed MAGLEV such as "Linimo" MAGLEV train which is operating in Japan since 2005 and the Korean MAGLEV "ECOBEE" which represent a demonstration line.

Let us consider a U-shaped core U-shaped track electromagnet as shown in Fig. 4.14, where the main geometrical parameters are identified. The study of such system goes through the prediction of the magnetic force which will be done based on the magnetic energy W_m.

Fig. 4.14 U-shaped track
with U-shaped core
electromagnet

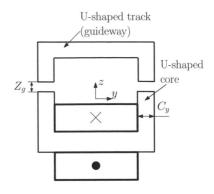

4.3.3.1 Magnetic Force

The magnetic force could be derived based on the magnetic energy W_m (Eq. (2.100))
where the energy could be expressed as derived in Chap. 2 (Eq. (2.85)) as:

$$W_m = \frac{1}{2} L i^2 \qquad (4.29)$$

where L and I are the coil electromagnet inductance and current.

Referring to Eq. (2.84), the inductance of the L could be expressed as:

$$L = N^2 P \qquad (4.30)$$

where P is the permeance of the closed flux loop and is expressed based on the
reluctance as, $P = \frac{1}{\mathcal{R}}$, and where N is the number of turn of the electromagnet coil.

Doing so, the prediction of the magnetic force goes through the determination of
the main flux loop electromagnet permeance, and thus through the expressions:

$$\begin{cases} F_z = F_{lev} = -\dfrac{I^2}{2} \dfrac{\partial L}{\partial z} \\ F_y = F_{gui} = -\dfrac{I^2}{2} \dfrac{\partial L}{\partial y} \end{cases} \qquad (4.31)$$

By neglecting the fringing effect, and when the track is facing the electromagnet
core as shown in Fig. 4.15a, the magnetic force produced by the U-shaped track
electromagnet is similar to the flat track one due to the identic air gap reluctance.
However, if we consider a displacement Δy along the y-axis of the electromagnet,
as illustrate in Fig. 4.15b, the fringing effect is no longer neglegible. Consequently,
the air gap permeance is greatly affected.

For the sake of simplicity, the permeances of the main flux loop will be limited
to the air gap ones. Indeed, giving the fact that the permeability of the iron parts is
higher than the air one, their permeances will be neglected. Hence, referring to the
geometrical parameters depicted in Fig. 4.15b, the considered permeance P could

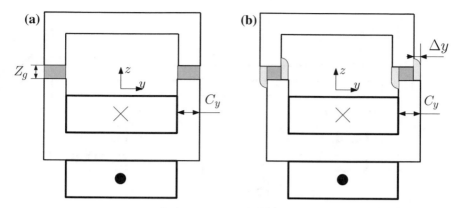

Fig. 4.15 U-shaped core U-shaped track electromagnet. **Legend**: a equilibrium position, b with disturbance along the y-axis

be expressed as:

$$P = \frac{\mu_0 L}{2} \left\{ \frac{C_y - \Delta y}{Z_g} + \frac{4}{\pi} \ln \left(1 + \frac{\pi \Delta y}{4 Z_g} \right) \right\} \tag{4.32}$$

where L is the active length of the electromagnet.

By substituting the corresponding permeance in Eq. (4.30), the inductance of the considered system will be expressed as:

$$L = \frac{\mu_0 L N^2}{2} \left\{ \frac{C_y - \Delta y}{Z_g} + \frac{4}{\pi} \ln \left(1 + \frac{\pi \Delta y}{4 Z_g} \right) \right\} \tag{4.33}$$

The application of the system (4.31) enables the determination of the leviation and guidance forces as:

$$\begin{cases} F_{lev} = \frac{\mu_0 N^2 I^2 L}{4} \left[\frac{C_y - \Delta y}{Z_g^2} + \frac{4 C_y}{4 Z_g^2 + \pi Z_g C_y} \right] \\ F_{gui} = \frac{\mu_0 N^2 I^2 L}{4} \left[\frac{1}{Z_g} - \frac{4}{4 Z_g + \pi C_y} \right] \end{cases} \tag{4.34}$$

Through the analysis of the levitation and guidance force expressions, one can deduce that they have nonlinear variation with respect to Δy, Z_g, and I. A linearized form of these equations around a given equilibrium point characterized by $(\Delta y_e = 0, Z_{ge}, I_e)$ is as follows:

$$\begin{cases} F_{lev} = C_{z_{lev}} (Z_g - Z_{ge}) + C_{I_{lev}} (I - I_e) \\ F_{gui} = C_{y_{gui}} (\Delta y - \Delta y_e) + C_{I_{gui}} (I - I_e) \end{cases} \tag{4.35}$$

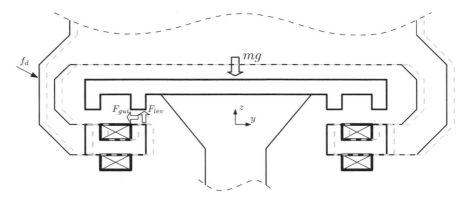

Fig. 4.16 Elementary system of MAGLEV train equipped by U-shaped track

where the constants $C_{z_{lev}}$, $C_{I_{lev}}$, $C_{y_{gui}}$, and $C_{I_{gui}}$ are defined as:

$$
\begin{cases}
C_{z_{lev}} = \dfrac{\partial F_{lev}}{\partial Z_g}\big|_{(Z_g=Z_{ge}, I=I_e)} \\[2mm]
C_{I_{lev}} = \dfrac{\partial F_{lev}}{\partial I}\big|_{(Z_g=Z_{ge}, I=I_e)} \\[2mm]
C_{y_{gui}} = \dfrac{\partial F_{gui}}{\partial \Delta y}\big|_{(\Delta y=0, I=I_e)} \\[2mm]
C_{I_{gui}} = \dfrac{\partial F_{gui}}{\partial I}\big|_{(\Delta y=0, I=I_e)}
\end{cases}
\tag{4.36}
$$

4.3.3.2 Elementary U-Shaped Track-Based EMS System

Let us consider an elementary system of MAGLEV train equipped by EMS system based on U-shaped track electromagnet as shown in Fig. 4.16.

The application of the *Newton*'s second law (Eq. (4.8)) leads to:

$$
\begin{cases}
m\ddot{z} = 2F_{lev} - mg - f_{dz} \\
m\ddot{y} = 2F_{gui} + f_{dy}
\end{cases}
\tag{4.37}
$$

where f_{dz} and f_{dy} are the z-axis and y-axis disturbance force f_d components, respectively.

At steady state, we have $\Delta y_e = 0$ and $mg = 2F_l|_{(Z_g=Z_{ge}, I=I_e)}$. Thus, the equilibrium point is characterized by:

$$
\begin{cases}
Z_g = Z_{ge} \\
\Delta y = \Delta y_e \\
I = I_e
\end{cases}
\tag{4.38}
$$

By considering the current-voltage expression derived in the previous section (Eq. 4.22) and the linearized form of the magnetic forces yielded by Eq. 4.36, the

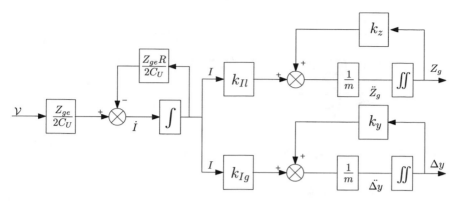

Fig. 4.17 Bloc diagram of the U-shaped track electromagnet based EMS system

state equations will be as follows:

$$
\begin{cases}
\ddot{z} = \dfrac{k_z}{m} Z_g + \dfrac{k_{Il}}{m} I \\[2mm]
\ddot{x} = \dfrac{k_y}{m} \Delta y + \dfrac{k_{Ig}}{m} I \\[2mm]
\dot{i} = \dfrac{Z_{ge}}{C_U} v - \dfrac{Z_{ge} R}{C_U} I
\end{cases}
\tag{4.39}
$$

where k_z, k_{Il}, k_y and k_{Ig} could be expressed based on $C_{z_{lev}}$, $C_{I_{lev}}$, $C_{y_{gui}}$, and $C_{I_{gui}}$ and where C_U is the constant characterizing the U-shaped track electromagnet and could be expressed based on the permeance P.

The corresponding bloc diagrams is illustrated in Fig. 4.17.

4.4 Electrodynamic Suspension Systems

4.4.1 Rectangular Coil Moving over Conducting Sheet

Let us consider a rectangular coil moving over a flat conducting sheet of conductivity σ with a constant speed u and at a constant hight h, as shown in Fig. 4.18 [16].

A cross section along the x-axis direction leads to the schematic view illustrated in Fig. 4.19. Firstly, only one filament will be considered with one directional current.

4.4.1.1 Magnetic Force

To simplify the study, we will consider only one direction current flow in the conducting sheet, that is $J = J_z \overrightarrow{z}$. Moreover, when the filament is moving over the sheet

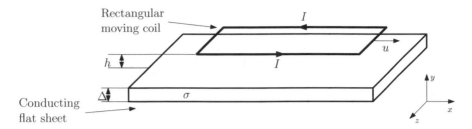

Fig. 4.18 Rectangular moving coil over a flat conducting sheet

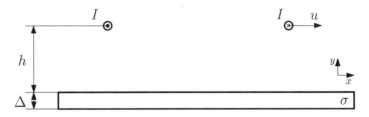

Fig. 4.19 Eddy currents diagram induced in conductive metal sheet

with a velocity ($u = u_x \vec{x}$), the current distribution in the track and its corresponding magnetic field have a steady behaviour with respect to the filament (i.e. $\partial B / \partial t = 0$). So, *Faraday*'s law becomes:

$$\begin{cases} \dfrac{\partial E_z}{\partial x} = 0 \\ \dfrac{\partial E_z}{\partial y} = 0 \end{cases} \tag{4.40}$$

Moreover, assuming an infinite length sheet, we found that $E_z = 0$.

Taking into account the previous assumptions, the application of *Ohm*'s law for a moving conductor ($J = \sigma(E + u \times B)$) leads to the determination of the current density in the sheet:

$$J_s = -\sigma u B_y \tag{4.41}$$

where the total magnetic field B_y is given by:

$$B_y = B_0 + B_1 \tag{4.42}$$

where B_0 and B_1 are the induced field by the filament and the field due to the current distribution in the sheet.

As the conducting sheet is characterized by a small thickness Δ, we could consider a uniform current density distribution along the sheet thickness. Thus, one can define the current density per unit length $K = J\Delta$. Thus, combining Eqs. (4.41) and (4.42), the longitudinal current density is expressed as:

$$K(x) = -\Delta \sigma u (B_0 + B_1) \tag{4.43}$$

The magnetic field B_1 is the field in the position x due to the current density $K(\eta)$ in the position η. Thus, based on the *Biot-Savart* law (Eq. (2.6), B_1 is given by:

$$B_1 = \int \frac{\mu_0}{2\pi} \frac{K(\eta)d\eta}{x - \eta} \tag{4.44}$$

By substituting B_1 in Eq. (4.43), one can deduce the differential equation in term of $K(x)$ as:

$$K(x) + \frac{u}{w\pi} \int \frac{K(\eta)d\eta}{x - \eta} = -\frac{2u}{w} \frac{B_0(x)}{\mu_0} \tag{4.45}$$

where w is the characteristic velocity and is as $w = \dfrac{2}{\mu_0 \sigma \Delta}$.

Using the *Fourrier* transform, the resolution of the differential equation leads to:

$$K(x) = -\frac{2}{\mu_0} \frac{u}{w} \frac{1}{1 + u^2/w^2} \left(B_{0y}(x) - \frac{u}{\pi w} \int \frac{B_{0y} d\eta}{x - \eta} \right) \tag{4.46}$$

Using the Maxwell equation ($\nabla . B = 0$), Eq. (4.47) turns to be:

$$K(x) = -\frac{2}{\mu_0} \frac{u}{w} \frac{1}{1 + u^2/w^2} \left(B_{0y}(x) - \frac{u}{w} B_{0x} \right) \tag{4.47}$$

where B_{0x} and B_{0y} are the components of the induced magnetic field by the filament and are expressed based on the *Biot-Savart* law as:

$$\begin{cases} B_{0x} = \dfrac{\mu_0 I}{2\pi} \dfrac{h}{x^2 + h^2} \\ B_{0y} = \dfrac{\mu_0 I}{2\pi} \dfrac{x}{x^2 + h^2} \end{cases} \tag{4.48}$$

Doing so, the forces on the moving filament are equal and opposite to the resultant forces on the sheet. Thus, using *Lorenz* force equation ($F = J \times B$), the magnetic force component are as follows:

$$\begin{cases} F_x = -\int K(x) B_{0y} \\ F_y = \int K(x) B_{0x} \\ F_z = 0 \end{cases} \tag{4.49}$$

By assuming that $h = x/c$ with $c = u/w$, the lift and drag forces are expressed as:

$$\begin{cases} F_l = F_y = \dfrac{\mu_0 I^2}{4\pi h} \dfrac{u^2}{w^2 + u^2} \\ F_d = F_x = \dfrac{\mu_0 I^2}{4\pi h} \dfrac{uw}{w^2 + u^2} \end{cases} \tag{4.50}$$

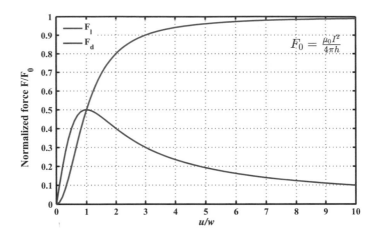

Fig. 4.20 Lift and drag forces versus normalized velocity (u/w) of the a current filament moving over a conducting sheet

The variation of the Lift and drag forces with respect to the normalized velocity (u/w) is shown in Fig. 4.20. From its analysis, one can deduce that low speed (*i.e.* $u/w < 1$) state is characterized by large drag with low lift. For critical speed (*i.e.* $u/w = 1$), the drag and lift forces are equal. At high speed when $u \gg w$, the lift force is widely larger than the drag one [17].

At standstill, the magnetic field of the current will fully permeate the sheet conductor. The magnetic field lines will be perfect circles about the current center. At very low speeds, ($u \ll w$) the field still permeates the sheet and the field lines will still be very near to the circular shape, as shown in Fig. 4.21a.

As the conductor speed is increased reaching approximately the characteristic velocity ($u \sim w$), the magnetic field still penetrate the conductive sheet but it loses it circular shape as the conductor leaves the region of magnetic field additional currents are induced to maintain the presence of the field as illustrated in Fig. 4.21b.

When the speed is increased to overcome the characteristics velocity, the conductivity of the sheet prevents the magnetic field from any significant penetration. The conductor is moving sufficiently fast that significant resistive dissipation does not occur. Each section of sheet generates the exact required current to perfectly shield the interior of the conductive sheet from the magnetic field. This situation is shown in Fig. 4.21c.

4.4.1.2 Elementary EDS-Based System

Let consider an EDS-based MAGLEV train. At the bottom on the train are installed rectangular superconducting (SC) coil which is moving over a conducting sheet track as shown in Fig. 4.22.

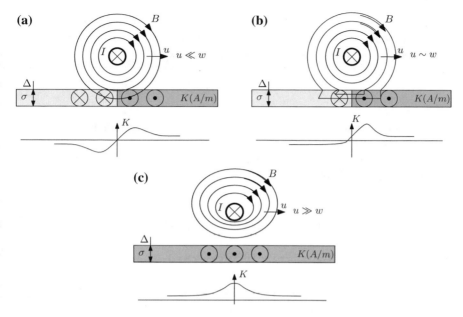

Fig. 4.21 Magnetic field and induced current distribution. **Legend**: **a** at low speed $u \ll w$, **b** at critical speed $u \sim w$, **c** at hight speed $u \gg w$

If we consider that the levitated car has a mass m, the application of the *Newton*'s second law leads to:

$$\begin{cases} m\ddot{y} = -mg + F_{lev} \\ m\ddot{x} = F_p - F_{dc} - f_{ad} \end{cases} \qquad (4.51)$$

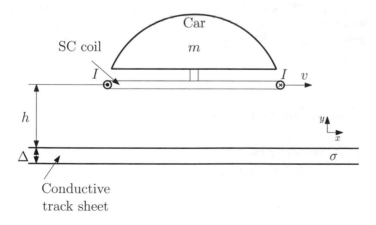

Fig. 4.22 EDS-based MAGLEV with SC coil moving over conductive sheet

where F_{lev} and F_{dc} are the levitation and drag forces exerted on the coil and where F_p is the propulsion force and f_{ad} is the aerodynamic drag force on the car.

Commonly, the aerodynamic drag force is expressed as:

$$f_{ad} = \frac{1}{2}\rho u^2 C_p A \tag{4.52}$$

where ρ is the density of the fluid (in this case the air), u in the speed, C_p is the drag coefficient, and A is the cross-section of the area.

As the SC coil current I is constant, the steady state point is characterized by (u_0, y_0), under which system (4.51) gives:

$$\begin{cases} F_{lev}(u_0, y_0) = mg \\ F_{dc}(u_0, y_0) = F_p - \frac{1}{2}\rho u_0^2 C_p A \end{cases} \tag{4.53}$$

The linearised equations of a moving coil may be written as:

$$\begin{cases} m\ddot{y} = k_{ul}u + k_{yl}y \\ m\ddot{x} = k_{ud}u + k_{yd}y \end{cases} \tag{4.54}$$

where the constants k_{ul}, k_y, k_{ud}, and k_y are yielded based on *Taylor* expansion.

The application of *Laplace* transform to such a system leads to the characteristic equation of the electrodynamic levitation system which is expressed as:

$$\left(s^2 - \frac{k_{yl}}{m}\right)\left(s - \frac{k_{ud}}{m}\right) - \frac{k_{ul}k_{yd}}{m^2} = 0 \tag{4.55}$$

The first term is related to vertical motion and second term horizontal (longitudinal) motion. This implies that the vertical oscillatory motion of the superconducting levitation system is almost uncoupled from its horizontal motion. Equation (4.55) reveals that the vertical dynamics of a superconducting levitation system are influenced only by the height y.

4.4.2   Figure-Eight Null-Flux Coil Based EDS System

Regarding EDS systems, a SC coil moving facing a figure-eight null-flux could be considered [18]. Such system is adopted and prototyped by the Japanese in the "L0" MAGLEV train [14].

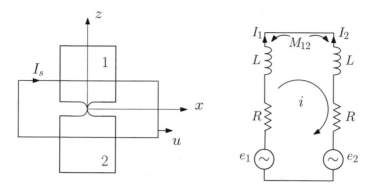

Fig. 4.23 Equivalent flux of a single SC coil interacting with a null-flux coil

4.4.2.1   Single Figure-Eight Null-Flux Coil

The simplest model is a single figure-eight null flux coil facing a moving SC coil as shown in Fig. 4.23, where each branch is characterized by a resistance R and an inductance L.

The coupling between the upper and lower loops is described by the mutual inductance M_{12}, and the coupling between the moving SC coil and the upper and lower loops of the null-flux coil is modeled by the mutual inductance M_{s1} and M_{s2}, respectively.

If we consider that the SC coil is moving only alon the x direction with a velocity v, the induced voltages in the upper (e_1) and lower (e_2) are as:

$$\begin{cases} e_1 = -I_s u \dfrac{\partial M_{s1}(x, y, z)}{\partial x} \\ e_2 = -I_s u \dfrac{\partial M_{s2}(x, y, z)}{\partial x} \end{cases} \tag{4.56}$$

Based on the equivalent circuit, the voltage equation is expressed as:

$$2Ri + 2(L - M_{12})\frac{di}{dt} = e_1 - e_2 \tag{4.57}$$

Giving the fact that the force $f_s = -\partial W/\partial s$, where W is the energy and s is the direction of the motion, the three component of the magnetic force could be expressed as:

$$\begin{cases} f_x = I_s i \left(\dfrac{\partial M_{s1}(x, y, z)}{\partial x} - \dfrac{\partial M_{s2}(x, y, z)}{\partial x} \right) \\ f_y = I_s i \left(\dfrac{\partial M_{s1}(x, y, z)}{\partial y} - \dfrac{\partial M_{s2}(x, y, z)}{\partial y} \right) \\ f_z = I_s i \left(\dfrac{\partial M_{s1}(x, y, z)}{\partial z} - \dfrac{\partial M_{s2}(x, y, z)}{\partial z} \right) \end{cases} \tag{4.58}$$

where i is the current in the equivalent circuit loop which is as $i = I_1 = -I_2$.

Basically, the mutual inductance between two conductors $c1$ and $c2$ could be derived using the *Neumann*'s formula as:

$$M = \frac{\mu_0}{4\pi} \int_{c1} \int_{c2} \frac{dxdx' + dydy' + dzdz'}{\sqrt{(x - x')^2 + (y - y')^2 + (z - z')^2}} \tag{4.59}$$

where x, y, and z and x', y', and z' are the variables of conductors $c1$ and $c2$, respectively.

Such formulation can be generally applied to any geometry of the conductors. However in some cases, analytical formulations are too complicated. Therefore, numerical procedures turn to be necessary. With this said, simplifying hypothesis could be adopted in order to to apply the analytical formulation. Accounting for the fact that a MAGLEV vehicle usually consists in group of SC coils, arranged in one row with alternating polarities, the mutual inductance between a group of SC coils and a null-flux coil could be expressed as:

$$M_{sj} = M_{pj}(y, z)cos(\frac{\pi}{\tau}x) \tag{4.60}$$

where M_{pj} is the peak value of the mutual inductance between one SC coil and the jth loop of the null-flux coil which depend only of the displacement along the z- and y- axis, and where τ is the pole pitch of the SC coil.

Based on the differential equation in terms of i (Eq. (4.57)) and considering Eqs. (4.56) and (4.60), the current i could be expressed as:

$$i = \frac{E_1 - E_2}{2\sqrt{R^2 + \omega^2(L + M_{12})^2}} \sin(\omega t - \varphi) \tag{4.61}$$

where:

$$\begin{cases} \varphi = \arctan\left(\frac{\omega(L - M_{12})}{R}\right) \\ E_j = \omega I_s M_{pj}, \quad j = 1, 2 \\ \omega = \frac{\pi}{\tau}u \end{cases} \tag{4.62}$$

4.4.2.2 A Pair of Figure-Eight Null-Flux Coil Without Cross Connection

Let us consider a pair of SC coils aboard of a MAGLEV vehicle interacting with figure-eight null-flux coils mounted on the two sides of the guideway as shown in Fig. 4.24, with i and i' are the currents in the 1–2 and 3–4 null-flux coils, respectively.

So, the three components of the magnetic forces acting on the vehicles are expressed as:

Fig. 4.24 Two SC coils interacting with two null-flux coil

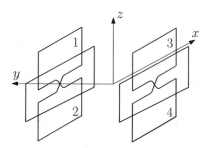

$$\begin{cases} F_x = I_s i \left(\dfrac{\partial M_{s1}(x, y, z)}{\partial x} - \dfrac{\partial M_{s2}(x, y, z)}{\partial x} \right) + I_s i' \left(\dfrac{\partial M_{s3}(x, y, z)}{\partial x} - \dfrac{\partial M_{s4}(x, y, z)}{\partial x} \right) \\[2mm] F_y = I_s i \left(\dfrac{\partial M_{s1}(x, y, z)}{\partial y_1} - \dfrac{\partial M_{s2}(x, y, z)}{\partial y_1} \right) + I_s i' \left(\dfrac{\partial M_{s3}(x, y, z)}{\partial y_2} - \dfrac{\partial M_{s4}(x, y, z)}{\partial y_2} \right) \\[2mm] F_z = I_s i \left(\dfrac{\partial M_{s1}(x, y, z)}{\partial z} - \dfrac{\partial M_{s2}(x, y, z)}{\partial z} \right) + I_s i' \left(\dfrac{\partial M_{s3}(x, y, z)}{\partial z} - \dfrac{\partial M_{s4}(x, y, z)}{\partial z} \right) \end{cases}$$
$$(4.63)$$

where y_1 and y_2 are the relative displacement along the y-axis of the left and right SC coils, respectively.

If we suppose that the vehicle has a displacement y, from the equilibrium position y_0, to the right side, y_1 and y_2 are expressed as: $y_1 = y_0 + y$ and $y_2 = y_0 - y$.

As the right and left null-flux coils are identical, we can assume that $M_{s1} = M_{s3}$, $M_{s2} = M_{s4}$, and $i = i'$. Thus the force components turns to be:

$$\begin{cases} F_x = 2I_s i \left(\dfrac{\partial M_{s1}(x, y, z)}{\partial x} - \dfrac{\partial M_{s2}(x, y, z)}{\partial x} \right) \\[2mm] F_y = 0 \\[2mm] F_z = 2I_s i \left(\dfrac{\partial M_{s1}(x, y, z)}{\partial z} - \dfrac{\partial M_{s2}(x, y, z)}{\partial z} \right) \end{cases}$$
$$(4.64)$$

System (4.64) reveals that the studied system does not achieve the guidance function as the force along the y-axis is null. However, such system is dedicated to the levitation via z-axis forces with a drag force (F_x).

The Substitution of Eqs. (4.60) and (4.61) in the system (4.64) leads to:

$$\begin{cases} F_x = - \left(\dfrac{(E_1 - E_2)^2}{u\sqrt{R^2 + \omega^2(L - M_{12})^2}} \right) \sin(\omega t) \sin(\omega t + \varphi) \\[2mm] F_z = I_s \left(\dfrac{(E_1 - E_2)}{\sqrt{R^2 + \omega^2(L - M_{12})^2}} \left(\dfrac{\partial M_{p1}(x, y, z)}{\partial z} - \dfrac{\partial M_{p2}(x, y, z)}{\partial z} \right) \right) \cos(\omega t) \sin(\omega t + \varphi) \end{cases}$$
$$(4.65)$$

4.4.2.3 A Pair of Figure-Eight Null-Flux Coil with Cross Connection

Now, we will consider that the right and left null-flux coils are cross-connected as shown in Fig. 4.25, where the corresponding equivalent circuit is illustrated. Giving

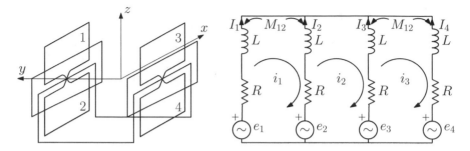

Fig. 4.25 Cross connected null-flux coils for suspension system with its equivalent circuit

Fig. 4.26 Simplified
equivalent circuit of the cross
connected null-flux coils

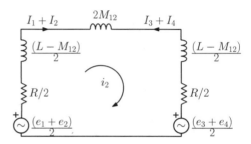

the fact that the left and side null-flux coils are taken away, their are mutual inductance
between them.

In the equivalent circuit represented in Fig. 4.25, the resistance and the inductance
of the cross-connecting cables are neglected and the i_2 is the current flowing the cross-
connecting cable. To simplify the study, a single loop circuit equivalent to the one
shown in Fig. 4.25 has been defined. This latter is illustrated in Fig. 4.26.

Based on the simplified equivalent circuit shown in Fig. 4.26, the voltage equation
could be rewritten as follows:

$$Ri_2 + (L + M_{12})\frac{di_2}{dt} = \frac{1}{2}[(e_1 + e_2) - (e_3 + e_4)] \qquad (4.66)$$

As it can be seen, the simplified equivalent circuit (Fig. 4.26) is similar to the
equivalent circuit of a single null-flux coil (Fig. 4.26). Thus, the current i_2 could be
expressed by the same way of the current i (Eq. (4.61)) as:

$$i_2 = \frac{(E_1 + E_2) - (E_3 + E_4)}{2\sqrt{R^2 + \omega^2(L + M_{12})^2}}\sin(\omega t - \varphi_2) \qquad (4.67)$$

where $\varphi_2 = \arctan\left(\frac{\omega(L + M_{12})}{R}\right)$.

Doing so, similar to the system (4.64), the magnetic force components due to the
cross-connection between the left-side and right-side null-flux coils are as:

$$\begin{cases} f_x = \frac{1}{2}I_s i_2 \left[\left(\frac{\partial M_{s1}}{\partial x} + \frac{\partial M_{s2}}{\partial x} \right) - \left(\frac{\partial M_{s3}}{\partial x} + \frac{\partial M_{s4}}{\partial x} \right) \right] \\ f_y = \frac{1}{2}I_s i_2 \left[\left(\frac{\partial M_{s1}}{\partial y_1} + \frac{\partial M_{s2}}{\partial y_1} \right) - \left(\frac{\partial M_{s3}}{\partial y_2} + \frac{\partial M_{s4}}{\partial y_2} \right) \right] \\ f_z = \frac{1}{2}I_s i_2 \left[\left(\frac{\partial M_{s1}}{\partial z} + \frac{\partial M_{s2}}{\partial z} \right) - \left(\frac{\partial M_{s3}}{\partial z} + \frac{\partial M_{s4}}{\partial z} \right) \right] \end{cases} \tag{4.68}$$

If we consider that:

$$\begin{cases} dy_1 = dy \\ dy_2 = -dy \\ \frac{\partial M_{s1}}{\partial y} = frac\partial M_{s3}\partial y \\ \frac{\partial M_{s2}}{\partial y} = \frac{\partial M_{s4}}{\partial y} \end{cases} \tag{4.69}$$

thus the force components turns to be as follows:

$$\begin{cases} f_x = \frac{1}{2}I_s i_2 \left[\left(\frac{\partial M_{s1}}{\partial x} + \frac{\partial M_{s2}}{\partial x} \right) - \left(\frac{\partial M_{s3}}{\partial x} + \frac{\partial M_{s4}}{\partial x} \right) \right] \\ f_y = I_s i_2 \left(\frac{\partial M_{s1}}{\partial y} + \frac{\partial M_{s2}}{\partial y} \right) \\ f_z = \frac{1}{2}I_s i_2 \left[\left(\frac{\partial M_{s1}}{\partial z} + \frac{\partial M_{s2}}{\partial z} \right) - \left(\frac{\partial M_{s3}}{\partial z} + \frac{\partial M_{s4}}{\partial z} \right) \right] \end{cases} \tag{4.70}$$

By considering an elementary EDS-based figure-eight null-flux coil system as illustrated in Fig. 4.27, the resultant of the magnetic force within the three directions (x, y, z) is the sum of:

- the force due to the interaction between the SC coils and the left-side and right-side null-flux coils when the two coils are not cross-connected (F_{sii}, $s = x, y, z; i = 1, 2$), and
- the force produced by these coils resulting from the cross-connection (F_{sij}, $s = x, y, z; i = 1, 2; j = 1, 2$).

Referring to the system (4.63), the force components due to the interaction between the SC coils and the left-side and right-side null-flux coils when the two coils are not cross-connected are as follows:

$$\begin{cases} F_{x11} + F_{x22} = -I_s^2 \frac{u u_{c1}}{u^2 + u_{c1}^2} \frac{\pi (M_{p1} - M_{p2})^2}{\tau (L - M_{12})} \\ F_{y11} + F_{y22} = 0 \\ F_{z11} + F_{z22} = -I_s^2 \frac{u^2}{u^2 + u_{c1}^2} \frac{M_{p1} - M_{p2}}{L - M_{12}} \left(\frac{\partial M_{p1}}{\partial z} - \frac{\partial M_{p2}}{\partial z} \right) \end{cases} \tag{4.71}$$

where u_{c1} is the characteristic speed associated to the vertical displacement which is $u_{c1} = \frac{\tau R}{\pi (L - M_{12})}$.

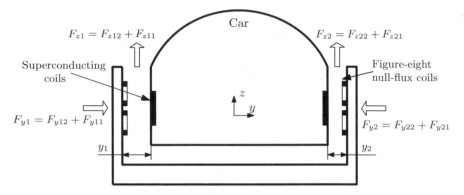

Fig. 4.27 Cross-section view of the null-flux suspension system

Considering the system (4.70), the force components produced by the cross-connection between the left-side and right-side null-flux coils could be expressed as:

$$
\begin{cases}
F_{x12} + F_{x21} = -\dfrac{I_s^2}{4} \dfrac{u u_{c2}}{u^2 + u_{c2}^2} \dfrac{\pi[(M_{p1} + M_{p2}) - (M_{p3} + M_{p4})]^2}{\tau(L + M_{12})} \\[2ex]
F_{y12} + F_{y21} = -\dfrac{I_s^2}{2} \dfrac{u^2}{u^2 + u_{c2}^2} \dfrac{(M_{p1} + M_{p2}) - (M_{p3} + M_{p4})}{L + M_{12}} \left(\dfrac{\partial M_{p1}}{\partial y} + \dfrac{\partial M_{p2}}{\partial y} \right) \\[2ex]
F_{z12} + F_{z21} = -\dfrac{I_s^2}{4} \dfrac{u^2}{u^2 + u_{c2}^2} \dfrac{(M_{p1} + M_{p2}) - (M_{p3} + M_{p4})}{L + M_{12}} \\[2ex]
\qquad\qquad\qquad \times \left[\left(\dfrac{\partial M_{p1}}{\partial z} + \dfrac{\partial M_{p2}}{\partial z} \right) - \left(\dfrac{\partial M_{p3}}{\partial z} + \dfrac{\partial M_{p4}}{\partial z} \right) \right]
\end{cases}
$$

$$(4.72)$$

where u_{c2} is the characteristic speed associated to the lateral displacement which is $u_{c2} = \dfrac{\tau R}{\pi(L + M_{12})}$.

4.5   Conclusion

This chapter is devoted to the magnetically-levitated trains which are more and more integrated in rail way systems. In order to highlight the historical back ground of such technology, a study statement has been carried out with emphasis on the over time progress of the MAGLEV trains. Moreover, the classification of such trains with respect to several criteria has been presented. Within this said, and giving the fact that the most innovative MAGLEV trains function is the levitation, their classification according to the suspension system has been deeply investigated. Thus, the electromagnetic suspension and electrodynamic suspension systems has been studied.

Basically, EMS systems are based on the attractive magnetic forces which are mainly the results of the interaction between the magnetic field produced by an electromagnet and a ferromagnetic sheet. Doing so, two electromagnet configurations are studied which are:

- flat track with U-shaped core electromagnet which achieve a mono-function the levitation or the guidance depending on its emplacement. Such concept is adopted for high speed MAGLEVs,
- U-shaped track with U-shaped core electromagnet which accomplish the levitation as well as the guidance. However, it is adequate only for low speed MAGLEVs.

Considering EDS technology and to achieve high force density, SC coils are required. Indeed, the interaction of such coils with a short-circuited electric circuit leads to the production of high repulsive magnetic forces. With this said, two systems have been studied:

- a moving SC coil over conducting sheet,
- a moving SC coil facing figure-eight null-flux coil.

The chapter has been achieved by a derivation of the produced magnetic forces and of the state equations of both EMS and EDS systems.

References

1. M. Dhingra, R.C. Sharma, M.H. Salmani, An introduction & overview to mangetically lavitated train. J. Sci. **5**(11), 1117–1124 (2015)
2. I. Boldea, Linear electromagnetic actuators and their control: a review. J. Eur. Power Electr. Driv. **14**(1), 43–50 (2004)
3. R. Hellinger, P. Mnich, Linear motor-powered transportation: history, present status, and future outlook. Proc. IEEE **97**(11), 1892–1900 (2009)
4. L. Yan, Development and application of the MAGLEV transportation system. IEEE Tran. Appl. Supercond. **18**(2), 92–99 (2008)
5. J. Meins, L. Miller, W.J. Mayer, The high speed MAGLEV transportation system transrapid. IEEE Trans. Mag. **24**(2), 808–811 (1988)
6. D.Y. Park, B.C. Shin, H. Han, Korea's urban MAGLEV program. Proc. IEEE **97**(11), 1886–1891 (2009)
7. S. Yamamura, Magnetic levitation technology of tracked vehicles present statusandprospects. IEEE Trans. Mag. **12**(6), 874–878 (1976)
8. I. Boldea, L. Tutelea, W. Xu, M. Pucci, Linear electric machines, drives and MAGLEVs: an overview. IEEE Trans. Ind. Electr. **65**(9), 7504–7515 (2018)
9. I. Boldea, *Linear Electric Machines, Drives, and MAGLEVs Handbook* (CRC Press, Taylor and Francis Group, New York, USA, 2013)
10. Z. Zhang, L. She, L. Zhang, C. Shang, W. Chang, Structural optimal design of a permanent-electromagnetic suspension magnet for middle-low-speed MAGLEV trains. IET Electr. Syst. Transp. **1**(2), 61–68 (2011)
11. H.-W. Lee, K.-C. Kim, J. Lee, Review of MAGLEV train technologies. IEEE Trans. Mag. **42**(7), 1917–1925 (2006)
12. Z. Zhang, L. She, L. Zhang, C. Shang, W. Chang, The inductrack: a simpler approach to magnetic levitation. IEEE Trans. Appl. Supercond. **10**(1), 901–904 (2000)

13. M. Mofushita, T. Azukizawa, S. Kanda, N. Tamura, T. Yokoyama, A New MAGLEV system for magnetically levitated carrier syetem. IEEE Trans. Veh. Technol. **38**(4), 230–236 (1989)
14. H.-S. Han, D.-S. Kim, *Magnetic Levitation: MAGLEV Technology and Applications* (Springer-Nature, New York, USA, 2016)
15. Z. Zhang, L. She, L. Zhang, C. Shang, W. Chang, Structural optimal design of a permanent-electro magnetic suspension magnet for middle-low-speed MAGLEV trains. IET Electr. Syst. Transp. **1**(2), 61–68 (2011)
16. F.C. Moon, *Magneto-solid mechanics* (Wiley, New York, USA, 1984)
17. A.R. Eastham, W.F. Hayes, MAGLEV systems developement status. *IEEE AES Mag.* (1988)
18. K.R. Davey, Designing with null flux coils. IEEE Trans. Mag. **33**(5), 4327–4334 (1977)

Printed in the United States
By Bookmasters